Statistical Geometry and Applications to Microphysics and Cosmology

Fundamental Theories of Physics

An International Book Series on The Fundamental Theories of Physics: Their Clarification, Development and Application

Volume 92

Statistical Geometry and Applications to Microphysics and Cosmology

by

Sisir Roy
Indian Statistical Institute,
Calcutta, India

KLUWER ACADEMIC PUBLISHERS
DORDRECHT / BOSTON / LONDON

A C.I.P. Catalogue record for this book is available from the Library of Congress.

ISBN 0-7923-4907-5

Published by Kluwer Academic Publishers,
P.O. Box 17, 3300 AA Dordrecht, The Netherlands.

Sold and distributed in the U.S.A. and Canada
by Kluwer Academic Publishers,
101 Philip Drive, Norwell, MA 02061, U.S.A.

In all other countries, sold and distributed
by Kluwer Academic Publishers,
P.O. Box 322, 3300 AH Dordrecht, The Netherlands.

Printed on acid-free paper

Printed in the Netherlands

"Perhaps the promising way to attack the local geometric problems would be with a combination of the two ideas....... the statistical approach and a geometry of lumps"

Karl Menger

Dedicated to the memory of **Professor Nathan Rosen**.

CONTENTS

PREFACE

This book is mainly about the development of the ideas of various physicists and mathematicians of this century on the stochastic nature of space-time. From the methodological point of view it is known as "**Ontological Geometry**". It covers roughly the period between 1930 and 1996. The earliest plan for this work came to my mind in on cold day of January 1986 when I was traveling from Calcutta, my home town, to Institut Henri Poincaré, Paris, on the invitation of Jean Pierre Vigier.

There are three main views on the nature of space and time :

- **The container view,**
- **The prime stuff**
- **The relational view.**

In this book, I have mainly discussed the developments of the relational view of space-time where space and time are not self-existing objects but a network of relations among factual items / things and their changes. The mathematical properties (like topological, affine, metrizability) assigned to space-time arise from the set of things, elementary particles, fields etc. and their changes. The recent results of high energy scattering and the theoretical developments of the string theory clearly indicate a revolutionary change in our understanding of the basic structure of space-time. I expounded the idea of C (cloud)- space in the spirit of Karl Menger mainly in the domain of Microphysics and Cosmology.

Menger introduced two main ideas to tackle geometric problems :

- The Geometry of hazy lumps
- Statistical approach.

This led Menger to introduce a distribution function rather than a definite number associated with every pair of elements in space. Here, the distinguishability between points got a probabilistic meaning. The hazy lumps may be thought of as seat of elementary particles or string like objects in contemporary physics. Mathematically speaking, the set-element relation acquired probabilistic meaning. In this way a statistical or probabilistic structure of metric space has been developed. In essence, Menger introduced the concept of probability into the foundations of geometry. Our usual metrics may be a kind of statistical average of stochastic or fluctuating metrics. So, space-time should be considered as derived concept. It raises some deep philosophical issues : in Leibniz's words

" Space is an order of possible co-existents and time an order of successives".

Now, the question arises as to how the ordering in its very generic sense is produced from a set of hazy lumps and their changes?

We proposed a kind of spontaneous fluctuation which gives rise to this kind of geometric order; for example, the order associated to the very adjective "topological". The concept of spontaneity and irreversibility occupied a central position in the history of science and philosophy since the birth of human civilization. The publication of **The Origin of Species** by Darwin and the microscopic interpretation of kinetic theory of gas by Boltzmann raised these issues to a new level of understanding and a unified picture is rapidly emerging.

For help in writing one needs critics (I certainly do) - the harsher and more unmerciful, the better. I owe a special debt to my wife Malabika (happens to be also a physicist) who literally tore apart the first version of first few chapters. She also tolerated apart much irascibility and postponement of research. One of her earlier comments began "you are vague....". I am indebted to Prof. Nathan Rosen for his constant inspiration and encouragement to write such a book but, unfortunately, he is no longer with us. I am also indebted to Prof. F. Guerra who inspired me on various occasions in this field. Menas Kafatos is greatly acknowledged for his critical remarks and editorial comments on the first chapter of this book. Kh. Namsrai, Silvio Bergia, J. Sanchez-Gomez, Myron W. Evans, L. de La Pêna, Y.Takano, Alfanso Rueda, R.K.Roychowdhury, R.L.Brahmachari and R.Ramanathan are also acknowledged for valuable comments and suggestions.

I am much grateful to Prof Jean Pierre Vigier who constantly encouraged me to develop the idea of fluctuating vacuum to understand quantum reality through invalauble discussions and criticism during my entire period of stay at Paris. English is not my native language. I believe that to some extent every language provides a different way of describing the common reality in which we are embedded. In this respect I am specially thankful to Swami Atmapriyananda (a physicist turned monk) at Ramkrishna Mission, Belurmath, West Benga, INDIA, for having inspired me to look into the above mentioned issues related to space even in the oldest literature in existence (like Vedas and Upanishad). This has helped me a lot to have a unified world view.

I wish also to acknowledge the continuous support in various forms from the present director, Prof. S.B.Rao of Indian Statistical Institute founded by late Prof. P.C.Mahalanobis, a man of deep insight and broad vision so as to write a book mainly devoted to theoretical physics even remaining within the purview of Statistical Institute. I owe a special debt to the chief editor Prof. A.Van der Merwe who corrected countless English solecism, improving clarity and tolerating my insistence on placing commas and periods after quotation marks for chapter 1. Finally I would like to thank Margaret Deignan of Kluwer Academic Publishers for her support and help in preparing some of the aspects of the preparation of the manuscript.

At last let me confess the very fact that when I started to write this book it seemed to me quite fascinating but a routine work. But as I was on my way to finish the manuscript, it started haunting my innerself endlessly searching for a new order:

" Almost falling Christopher at last reaches the bank, and he says to the child : Here we are! How thou wert! Child, who art thou? And the child answers : I am the day soon to be born".

Sisir Roy

Calcutta

INTRODUCTION AND HISTORICAL SURVEY

"One reason why mathematics enjoy esteem, above all of the science, is that its propositions are absolutely certain and indisputable, while those of all other sciences and to some extent debatable and in constant danger of being overthrown by newly discovered facts .... At this point one enigma presents itself ... How can it be that mathematics, being afterall a product of human thought which is independent of experience, is so admirably appropriate to the object of reality ? ... as far as the propositions of mathematics refer to reality, they are not certain; and as far as they are certain, they do not refer to reality"

Albert Einstein : Geometry and Experience.

1. INTRODUCTION AND HISTORICAL SURVEY:

1.1. Introduction :

In 19th century, Riemann [Riemann 1867] wrote a small but epoch-making paper "Über die Hypothesen - welche der Geometrie zu Grunde liegen" during his investigation to find the simplest facts from which the metrical relations of space may be derivable. He concludes this paper by saying :

"The question of the validity of the hypothesis of geometry in the infinitely small is bound up with the question of the ground of the metric relations of space. In this last question, which we may still regard as belonging to the doctrine of space, is found the application of the remark made above; that in a discrete manifoldness, the ground of its metric relations is given in the notion of it, while in a continuous manifoldness, this ground must come from outside. Either therefore the reality which underlies space must from a discrete manifoldness, or we must seek the ground of its metric relations outside it, in binding forces which act upon it.

The answer to these questions can only be got by starting from the conception of phenomena which has hitherto been justified by experience, and which Newton assumed as a foundation, and by making in this conception the successive changes required by facts which it cannot explain. Researches starting from general notions like the investigation we have just made, can only be useful in preventing this work from being hampered by too narrow views, and progress in knowledge of the interdependence of things from being checked by traditional prejudices.

This leads us into the domain of another science, of physics, into which the object of this work does not allow us to go today."

It is important to note that almost a century later, the developments of high-energy particle interactions lead us towards a deeper understanding of the structure of space-time at small distances. The new experiments are designed to probe distances up to 10^{-17} cm at various laboratories [Bracci 1983]. The existence of such a small distance was considered in two different contexts in elementary particle interactions. Some authors like Heisenberg [Heisenberg 1936], Born [Born 1938], Pais [Pais 1950], Ulhenbeck [Ulhenbech 1930], Yukawa [Yukawa 1950], Takano [Takano 1967], Markov [Markov 1984], Tamm [Tamm 1965], Kirzhnits [Kirzhnits 1966], Blokhintsev et al. [Blokhintsev 1973] introduced the concept of fundamental length in physics where this length is a universal constant. On the other hand, several authors like Efimov et al. [1963] considered this length not as a universal constant but that which characterizes only the domain of non-local interaction (or the size of extended particles, say, electrons) of the considered quantum fields. In fact, this is the basis of Efimov's constructions of a non-local theory. Efimov et al. [Efimov 1963] defined non-local field as the field obtained by averaging out the space-time $R^4(x)$. In a recent monograph, Namsrai [Namsrai 1986] analysed this situation in details, where earlier references concerning this problem are cited.

The recent theoretical and experimental success of high energy physics as well as the results in E-P-R-Bohm type experiments [Bohm 1960] and Neutron Interferometry [Rauch 1978] give a strong stimulus to the investigation of the space-time structure at small distances. The works of Wheeler et al. [Wheeler 1964] on super space, pre geometry and inflationary scenario at cosmological scale also raise a lot of interest regarding the fluctuation of space-time metric at small distances. In other words, the geometry is not deterministic, even though it looks so at the everyday scale of observation. The quantum fluctuations in the geometry of space are so great at small distances that even the topology itself fluctuates, makes "worms" and traps lines of force. So the idea of the stochastic character of space-time at small distance might play a great role in physical theory. The recent developments in String theory [Witten 1966] suggests that even the concept of metrizability might be lost at small distances.

We can briefly summarize the above results in the following way. The recent high energy experiments reveal that a minimal length (or fundamental length) may exist below which no measurement is possible. So the spatial continuum may have to be replaced by a granular or quantized space with a fundamental length built into it. But this hypothesis does not solve Poincaré dilemma of distinguishing between a transitive mathematical and an intransitive physical relation of equality. Now if we consider the geometry where points are not the primary entities but rather lumps and introduce the concept of probability so that the same two objects are sometimes treated as identical and sometimes as distinguishable , then it is possible to construct an alternate geometry which is more compatible with the physical reality. In this situation, if the space is considered to be a set of "hazy" or "fuzzy" lumps (known as cloud (C-) spaces) which do not have sharp boundaries, then we might get a consistent picture of geometry conforming with the physical reality. These lumps may be the seat of elementary particles or the size of the string.

In this new geometry we have introduced two basic notions :

- Concept of "hazy" or "fuzzy" lumps
- The statistical notion.

Here, the statistical element may arise due to the existence of random zero point field in the Universe or stochastic nature of string like objects. This geometry may be called as **Ontological Geometry** [Bunge 1977] where space and time are not self existing (absolute) objects of uncertain ontological status (neither things nor properties of things). Here, space-time should be considered as the **basic structure of all possible facts.**

During the developments of the physical theories at microscopic level as well as at cosmological scale, Menger [Menger 1951] and his collaborators developed the theory of probabilistic metric spaces from the mathematical point of view. In a paper entitled "Modern geometry and the theory of relativity" published in the book *Albert Einstein, Philosopher-Scientist* :

Karl Menger wrote : "Poincaré, in several of his famous essays on philosophy of science, characterized the difference between mathematics and physics as follows: In mathematics, if quantity A is equal to the quantity B and B is equal to C, then A is equal to C; that is, in modern terminology, mathematical equality is a transitive relation. But in the observable physical continuum "equal" means indistinguishable; and in this continuum, if A is equal to B, and B is equal to C, it by no means follow that A is equal to C ... That is to say, physical equality is not a transitive relation. Is this reasoning cogent? It is indeed easy to devise experiments which prove that the question whether two physical quantities are distinguishable, cannot always be answered by a simple Yes or No ... Elaboration of this idea leads to the concept of a space in which a distribution function rather than a definitive number is associated with every pair of elements. The number associated with two points of a metric space is called the distance between two points ... Such a statistical generalization of metric spaces appears to be well adapted for the investigation of physical quantities and physiological thresholds."

In 1983, Schweizer and Sklar [Schweizer 1983] published a book on probabilistic metric spaces which is essentially devoted to the study of structure of different type of probabilistic metric spaces as well as their topological properties. Virtually in all the mathematical models (as mentioned by Schweizer and Sklar) built to describe the various situations in physics, Psychometrics [Steven 1959], Communication theory [Shannon 1948] and Pattern recognition [Duda 1973] etc., it is assumed that the measurements are made with respect to a rigid reference frame. Remarks to the effect that this assumption may be unsatisfactory and that some of the uncertainties should be built into the geometry are scattered here and there in the literature ([Poincaré 1905, 1913]; [Hjelmslev 1923]; [March 1932]; [de Broglie 1935]; [Black 1937]; [Weyl 1952]; [Born 1955]; [Oppenheimer 1962]) along with suggestions on the proper way of doing them ([Penrose and Maccullam 1973]; [Penrose 1975]). There are also some serious attempts in this direction ([Eddington 1953]; [Rosen 1947]; [Blokhintsev 1973]; [Frenkel 1977]; [Takano 1961]; [Ingraham 1967]; [Fredericks 1976]; [Roy 1979]; [Vigier 1982]).

The aim of the present chapter is to present a systematic exposition of the works of various authors so as to understand the importance of studying Statistical Geometry both in Microphysics and in Cosmology.

1.2. Beginnings :

It was Einstein who first speculated in 1924 about fluctuations of the space-time metric. As far as the author's knowledge is concerned, Einstein did not elaborate this idea further. Louis de Broglie [de Broglie 1935], March [March 1937, 1938], tried to develop the idea of Stochastic space-time. March considered granular space-time and tried to construct a random metric in Minkowskian framework. But the idea was not developed rigorously and it required further physical insights. We had to wait for Menger to introduce the idea of Statistical Geometry in 1942 [Menger 1942] with full mathematical rigor, which was pursued and elaborated by his collaborators in the later decades. But it was Rosen who elaborated the idea of March with physical justification so as to make it compatible with the behavior of microparticles. Rosen's physical model has been shown to be justified by the work of Schweizer and Sklar (collaborators of Menger) within the framework of Menger's Statistical Geometry. So, at first we shall briefly discuss the works of Menger.

1.3. Menger, 1942 :

In 1942, Menger first proposed a probabilistic generalization of the metric spaces as developed by Frechet [Frechet 1906] and Hausdorff [Hausdorff 1914]. According to Menger [Menger 1951], the notion of distance in metric-space is essentially based on the introduction of a function d that assigns a non-negative real number $d(p,q)$ (distance between p and q) to every pair (p,q) of elements (points) of a non-empty set S. This function is assumed to satisfy the following conditions :

$$\begin{array}{llcll} (i) & d(p,q) & = & 0 \text{ if } p \text{ and } q \text{ coincide}, & \\ (ii) & d(p,q) & > & 0 \text{ if } p \text{ and } q \text{ are distinct}, & \\ (iii) & d(p,q) & = & d(p,q) \text{ for all } p,q \text{ in } S, & \\ (iv) & d(p,r) & \leq & d(p,q) + d(q,r) \text{ for all } p,q,r \text{ in } S. & \end{array} \tag{1.1}$$

Any function d satisfying [1(i)-1(iv)] is a metric on S. A metric space is a pair (S,d) where S is a set and d is a metric on S. Menger replaced the number $d(p,q)$ by a real function F_{pq} whose value $F_{pq}(x)$ for any real x, is interpreted as probability that the distance between p and q is less than x. Since probabilities can neither be negative nor be greater than 1, we have

$$0 \leq F_{pq}(x) \leq 1 \tag{1.2}$$

for every real x, and we have

$$F_{pq}(x) \leq F_{pq}(y) \tag{1.3}$$

for all values of x. Hence F_{pq} is a probability distribution function . Since distance cannot be negative, we have

$$F_{pq}(0) = 0. \tag{1.4}$$

Similarly :

(i) if $p = q$, then

$$F_{pq}(x) = 1, \quad \text{for all } x > 0. \tag{1.5}$$

(ii) if $p \neq q$, then

$$F_{pq}(x) < 1, \quad \text{for some } x > 0. \tag{1.6}$$

(iii) finally

$$F_{pq} = F_{qp}. \tag{1.7}$$

Here E.q. (I.i-I.iv) is essentially the triangle inequality. Menger [Menger 1951] and Wald [Wald 1943, 1955] proposed different generalizations of this triangle inequality. According to Menger's paper, the inequality reads

$$F_{pr}(x+y) \geq T\ (F_{pq}(x),\ F_{qr}(y)), \tag{1.8}$$

where for all p, q, r in S and for all real numbers x, y,

$$\begin{array}{llll} (i) & T(a,b) & = & T(b,a), \\ (ii) & T(a,b) & \leq & T(c,d), \text{ whenever } a \leq c, b \leq d, \\ (iii) & T(a,1) & > & 0, \text{ for } a > 0 \text{ and } T(1,1) = 1. \end{array} \tag{1.9}$$

Given (1.9:(i-iii)), the inequality (I.8) implies that our knowledge of the third sides of a triangle depends in a systematic manner on our knowledge of the other two sides. Menger also studied the notion of betweeness in this framework. Of course, generally, it is not all like the properties of ordinary metric betweenness.

1.4. Rosen, 1947 :

In 1947, Nathan Rosen initiated a discussion on the model of elementary particles and Statistical Geometry. His approach is related to that of March [March 1934, 1937, 1938]. Here, one will not obtain an exact value, say, of the x coordinate of a point because there does not exist an infinitesimally small measuring rod in nature. In a sense it indicates that there exists a smallest measurable distance as proposed by Snyder [Snyder 1947]; then it is reasonable to expect that repeated measurements will give values distributed about the mean value in a normal, or Gaussian distribution. Thus it was assumed that the probability of getting a value lying between ξ and $\xi + d\xi$ is given by

$$(2\pi a^2)^{-1/2} \int \exp[-(\xi - x)^2/2a^2] d\xi \tag{1.10}$$

where x is the mean value and a be a constant. This constant a can be related to the smallest measurable distance or fundamental length. For example, if one measures the position of an electron at a certain instant of time, the probability of obtaining values in the ranges $d\xi$, $d\zeta$, $d\rho$ near ξ, ζ, ρ is given by

$$d\omega = \chi(|\xi - x|, |\zeta - y|, |\rho - z|)d\xi d\zeta d\rho, \tag{1.11}$$

where

$$\chi(u, v, w) = (2\pi a^2)^{-\frac{1}{2}} \exp[-\frac{(u^2 + v^2 + w^2)}{2a^2}] \tag{1.12}$$

so that

$$\int_{-\infty}^{+\infty} \int_{-\infty}^{+\infty} \int_{-\infty}^{+\infty} \chi(u, v, \omega) du dv d\omega = 1 \tag{1.13}$$

for random variables u, v, ω. In this way Rosen arrived at the idea of a geometry with small regions instead of points. This elementary volume differs from the type considered by March in that it does not have definite boundary.

These small fuzzy regions or small clouds may be considered as the seat of elementary particles. Then two spaces can be constructed, an "abstract" space consisting of points and "observable" space in which one deals with these fuzzy regions correlated to the points of the former by means of a statistical distribution function in the form of a three-dimensional Gaussian error function. But such a type of function is not Lorentz invariant. One can obtain Lorentz covariance in the observable space by carrying out the usual Lorentz transformation in the abstract space. The usual Lorentz transformation is assumed to hold in the transformation from the abstract space of one reference frame to that of the other frame moving relative to it with uniform velocity. The transformation of a physical quantity in the observable space can be determined from the corresponding transformation in the abstract space if one knows how to go from the abstract to the observable space.

But the relativistic considerations posses the following difficulties : We have so far considered the uncertainty in the measurement of a space coordinate at a particular instant of time. Now the question is whether one should also introduce the uncertainty in the time as has been done for spatial coordinates. This means that we have an additional factor

$$(C/\sqrt{2\pi b^2}) \exp[-C^2(\tau - t)^2/2b^2] d\tau$$

in the former three-dimensional Gaussian distribution function for spatial variables. It immediately raises some difficulties from the operational points of view. In a three-dimensional distribution of spatial coordinates, it represents the probability distribution of measured values of the coordinate of a point, such as, position of an electron at a given instant of time. Here, time serves as a parameter to identify what is to be measured. But in a four-dimensional treatment, no such parameter is known with respect to which the measurement could be repeated. Rosen tried to solve it by considering the space-like hypersurfaces instead of dealing the whole four-dimensional space-time continuum.

Subsequently, Ingraham [Ingraham 1962, 1967] discussed the whole issue in a more elaborate fashion so as to have a consistent framework. Before elaborating the later proposal we shall concentrate our discussions on mathematical developments of distribution spaces within Menger framework. It will clearly establish Rosen's hypothesis of stochastic space-time on a rigorous mathematical foundation. It may be mentioned that though Rosen's idea was before this mathematical developments it was quite unnoticed for a long time. Schweizer and Sklar constructed distribution generated spaces and hence C-space (cloud - spaces) during their investigation to overcome the difficulties raised by Wald within Menger framework. It appears that neither Rosen nor Schweizer and Sklar were aware of each other's developments, though they happened to be of almost the same period.

1.5. Mathematical Developments 1956 - 1962 :

Immediately after the publication of the Menger's paper, Wald published a paper [Wald 1943] stating that the inequality (1.8):

"has the drawback that it involves an unspecified function $T(a,b)$ and one can hardly find sufficient justification for particular choice of this function. Furthermore, the notion of 'between' introduced by Menger on the basis of inequality (1.8) has the properties of the between relationship in metric-spaces only under restrictive conditions on the distribution functions" ....

Wald replaced the inequality (1.8) by the inequality

$$F_{pq}(x) \geq (F_{pq} * F_{qr})(x),$$

where the asterisk denotes the convolution, so that

$$(F_{pq} * F_{qr})(x) = \int_{\infty}^{\infty} F_{pq}(x-y)dF_{qr}(y) = \int_{0}^{x} F_{pq}(x-y)dF_{qr}(y)$$

On the basis of the above inequality, Wald introduced a notion of betweenness which has all the properties of ordinary betweenness.

Schweizer and Sklar started their collaboration in 1956 on probabilistic metric spaces and were looking at the difficulties raised by Wald. In 1957, they were able to avoid the difficulties in the course of studying certain particular spaces. It was shown that if T is taken to be any of the following functions :

$$\begin{aligned} W(a,b) &= max(a+b-1,0)\Pi(a,b) = abM(a,b) = min(a,b)M^*(a,b) \\ &= max(a,b)\Pi^*(a,b) = a+b-abW^*(a,b) = min(a+b,1) \end{aligned} \tag{1.14}$$

then the inequality (1.8) not only tells us that the probability $F_{pr}(x+y)$ depends in some monotonic way on the probabilities $F_{pq}(x)$ and $F_{qr}(y)$, but it also makes the nature of this dependence precise. In 1962, they studied a class of spaces called distribution generated spaces, which are very close to the physical spaces considered by Rosen et al. The basic idea is as follows:

Let S be a set. With each point p of S, associate an n-dimensional distribution function G_p whose margins are in $\mathcal{D}$, and with each pair (p,q) of distinct points of S, associate a $2n$-dimensional distribution function H_{pq} such that

$$H_{pq} = (\vec{u}, (\infty, \ldots, \infty)) = G_p(\vec{u})$$

$$H_{pq} = ((\infty, \ldots, \infty), \vec{v}) = G_q(\vec{v})$$

for any $\vec{u} = (u_1, \ldots, u_n)$ and $\vec{v} = (v_1, \ldots, v_n)$.

Furthermore, in R^n for any $n \geq 0$, let $Z(x)$ be the cylinder in R^{2n} given by

$$Z(x) = (\vec{u}, \vec{v}) \;\; in \;\; R^{2n}, \quad \|\vec{u} - \vec{v}\| < x,$$

and define F_{pq} in $\mathcal{D}^+$:

$$F_{pq}(x) = \int_{Z(x)} dH_{pq} = P_{H_{pq}} Z(x). \tag{1.15}$$

If we think of elements of S as "particles" then, for any Borel set A in R^n, the integral $\int_A dG_p$ is naturally interpreted as the probability that the particle p is in the set A and $F_{pq}(x)$ as the probability that the distance between the particles p and q is less than x. They call it as distribution generated spaces. Now if p is a point in a distribution generated space over R^n such that G_p is absolutely continuous, then any density g_p of G_p may be visualized as a "cloud" in R^n, a cloud whose density at any point of R^n measures the relative likelihood of finding the particle p in the vicinity of that point. In this way C-spaces or cloud spaces can be constructed. These

C-spaces play important role in describing the behavior of microparticles and in probing the structure of space-time at small distances.

Normal C-spaces: A C-space over R^n is normal if the density g_p of a non-singular point is spherically symmetric normal density i.e. if there is a number $\sigma > 0$ such that

$$g_p(\vec{u}) = (2\pi\sigma_p{}^2)^{-\frac{n}{2}} \exp[- \mid \vec{u} - \vec{c}_p \mid /2\sigma_p^2] \tag{1.16}$$

for all $\vec{u}$ in R^n. If p is singular, then we set $\sigma = 0$. Now if we take two distinct points p and q which are distinct and not both singular, then g_{pq} is also a normal, given by as follows:

$$g_{pq}(\vec{u}) = (2\pi\sigma_{pq}{}^2)^{-\frac{n}{2}} \exp(\mid \vec{u} \mid^2 /2\sigma_{pq}^2), \tag{1.17}$$

where

$$\sigma_{pq}{}^2 = \sigma_p{}^2 + \sigma_q{}^2 \tag{1.18}$$

This function $g_{pq}(\vec{u})$ has strikingly similar properties which Rosen considered in his model of elementary particles and statistical geometry.

1.6. Blokhintsev, Takano 1960 :

In 1961, Takano [Takano 1961] proved that the indefinite metric in Hilbert space can be attributed to the fluctuation of space-time. However, in subsequent publications, Takano discussed fluctuations of space-time in connection with the internal degrees of freedom of elementary particles. These internal degrees of freedom have been studied by Yukawa [Yukawa 1950] in his framework of non-local field theory and the structure of elementary particles. Takano speculated that these extra degrees of freedom are inherently random and hence give rise to the fluctuation of space-time metric. The present author along with Asanov et al. [Asanov 1988] have shown that Takano's speculation is generally true in case of Finslerian manifold. Hence Finslerian geometry is inherently probabilistic in nature.

In Takano's hypothesis, space-time is assumed to fluctuate or to be distributed probabilistically. Here, the fluctuation of space-time is regarded as a subnuclear concept and it's width is related to the so called universal length. As an example, Takano considered that the fluctuation of space-time will rise in the fifth dimension. The fifth dimension is a probability space as well as a metric space. In other words, the fluctuating four dimensional space is structurally the five-dimensional probability distribution of four dimensional subspaces corresponding to various values of ξ. The probability distribution function $\pi(\xi)$ is normalized to unity and put equal to the absolute square of the probability amplitude $\phi(\xi)$, which may be so called internal wave function of elementary particles. The metric fundamental form τ' can be related to that in four-dimensional Minkowski space (depicted by τ) as

$$\tau\prime^2 = \tau^2 - (\xi_2 - \xi_1)^2 \tag{1.19}$$

where

$$\tau^2 = c^2(t_2 - t_1)^2 - (x_2 - x_1)^2 - (y_2 - y_1)^2 - (z_2 - z_1)^2.$$

He calculated also the average propagator for the fields in this fluctuating space-time. In subsequent publications, this has been generalized so that the fluctuation of the space-time is assumed to be three-dimensional harmonic oscillation. Then the fundamental form τ' can be written as

$$\tau'^2 = \tau^2 - ({\rho_1}^2 + {\rho_2}^2 - 2\rho_1\rho_2 \cos\theta), \tag{1.20}$$

in which, ρ_1 and ρ_2 are the radii and θ is the angle between them. Here, the second term will play the part of $(\xi_2 - \xi_1)^2$ in the preceding definition. In this framework, for massless particles, the Coulomb potential will be modified after averaging over the probability distribution as

$$V_c' = (\sqrt{2}k_0/\sqrt{\pi})(\frac{k_0^2r^2}{4})\exp(\frac{k_0^2r^2}{4}) \times [k_1(\frac{k_0^2r^2}{4}) - k_2(\frac{k_0^2r^2}{4})] \tag{1.21}$$

which is not singular and equal to $\sqrt{2}k_0/\sqrt{\pi}$ at the origin and behaves like $1/r$ at infinity. In Takano's framework, the fluctuation of space-time may be assumed to rise not in the other dimension but in the four-dimensional space-time itself. This theory is reduced to the usual one in the limit of distribution of δ-function or no fluctuation.

The divergence problem in quantum field theory has also been studied by Takano in this framework. It gave rise to new insights to avoid divergency. The fluctuation of space-time as considered by Takano has a similarity with the random metrics considered and developed by mathematicians. Blokhintsev [Blokhintsev 1973] in the former Soviet Union also started his investigation on stochastic spaces almost at the same time of that of Takano. In his framework, stochastic spaces are spaces whose metrics depend on random quantities such as parameters, functions or quantum operators.

Essentially this idea is similar to that proposed by Takano i.e., random metrics. But Blokhintsev studied the mathematical and physical stochastic spaces in a more systematic manner. He considered the stochastic spaces and Einstein's equation for the gravitational field. It has deeper implications in the researches on gravitation and cosmology.

Blokhintsev defined stochastic space by considering the idea of "attributed spaces". The mapping of the attributes ξ on the coordinates x contains the parameter ζ, which takes on the random variables

$$x = X(\xi \mid \zeta). \tag{1.22}$$

The distribution of the possible values of the parameter ζ are given by the normalized probability

$$\int d\omega(\zeta) = 1. \tag{1.23}$$

where the mean value over the region of random parameters ζ can be written as

$$\bar{x} = < x > = \int X(\xi \mid \zeta) d\omega(\zeta). \tag{1.24}$$

The space $R^4(x)$ of the variable x is called stochastic space and $R^4(\xi)$ as attributed space. Again, if the deviation $(\bar{x} - x)$ is small, then,

$$x = \bar{x} - \Delta x(\bar{x}, \xi). \tag{1.25}$$

This relation is usually a mapping of the true coordinates x on to the space $R^4(x)$ of its mean values. The metric tensor in the space $R^4(x)$ can be written as

$$ds^2 = G_{\mu\nu}(x, \xi) dx^\mu dx^\nu. \tag{1.26}$$

The metric tensor $G_{\mu\nu}$ in $R^4(x)$ is a stochastic metric due to the presence of random parameters ξ. The metric induced by interacting fields may be stochastic. Several examples of such fields are considered by Blokhintsev. For example, turbulent motion of matter can also generate a stochasticity in the metric. In this case, the energy tensor of matter $T^{\mu\nu}$ can be expressed as

$$T^{\mu\nu} = T^{\mu\nu}(x, \xi). \tag{1.27}$$

Here, the random parameters ξ can be the initial values of the Lagrangian coordinates of the particle. From the Einstein's equation

$$R_{\mu\nu} - \frac{1}{2} g_{\mu\nu} R = 8\pi k c^4 T_{\mu\nu}(x, \xi) \tag{1.27}$$

it follows that metric tensor $g_{\mu\nu}(x)$ is also stochastic :

$$g_{\mu\nu} = g_{\mu\nu}(x, \xi) \tag{1.28}$$

If the fluctuations in the turbulent motion of matter are small in comparison with the mean values, then

$$T^{\mu\nu} = \bar{T}^{\mu\nu} + \delta T^{\mu\nu}, \tag{1.29}$$

where $T^{\mu\nu}$ describes the global motion of matter characterized by large scales L and period of T, whereas the term $\delta T^{\mu\nu}$ is due to turbulent motion of matter characterized by small scales and short periods $\tau(\lambda << L, \tau << T)$. The mean values of $\delta T^{\mu\nu}$ for time intervals comparable with T is assumed to be zero, and hence

$$\langle T^{\mu\nu} \rangle = \bar{T}^{\mu\nu}, \quad \langle \delta T^{\mu\nu} \rangle = 0. \tag{1.30}$$

The angular brackets denote averaging over the turbulent motion. So the metric tensor may be decomposed into two parts :

$$g_{\mu\nu} = \bar{g}_{\mu\nu} + \delta g_{\mu\nu}. \tag{1.31}$$

The magnitude of the turbulent fluctuation of matter $\delta T^{\mu\nu}$ is assumed to be small, and thus the quantities $\delta g_{\mu\nu}$ is also small compared with $\bar{g}_{\mu\nu}$, i.e., the mean values of $g_{\mu\nu}$. The length scale of fluctuation has been estimated to be of the order of 10^{-32} cm, which is much larger than the gravitational radii of particles. Snyder space has been shown to be a stochastic space whose metrics are determined by operators.

1.7. Random Metrics :

During the developments of random metrics by Takano, Blokhintsev et al., mathematicians independently investigated the structure of random metric space instead of defining it as probabilistic metric space. Their motivation was to analyze the probabilistic metric spaces within the standard measure theoretical model of probability theory. Spacek [Spacek 1956], Serstnev [Serstnev 1967], and Stevens [Stevens 1965] did some important works in this direction. In 1969, Sherwood [Schweizer 1983] approached it from a direction which has some similarity with the random metrics considered by Takano and Blokhintsev et al.

Sherwood introduced the concept of E-space. The points of an E-space are functions from a probability space (Ω, a, P) into a metric space (M, d). For each pair of functions in the space, the function $d(q, p)$ can be defined by

$$(d(p,q))(\omega) = d(p(\omega), q(\omega)) \quad \textit{for all } \omega \textit{ in } \Omega; \tag{1.32}$$

It is assumed to be a random variable on (Ω, a, P). The function F_{pq} is taken to be distribution function of this random variable , so that, for any real x,

$$F_{pq} = P(\omega \textit{ in } \Omega \mid (d(p,q))(\omega) \mid< x). \tag{1.33}$$

Thus F_{pq} is the probability that the distance between p and q is less than x. It has been shown that every E-space is a Menger space, and this is very closely related to distribution generated spaces as studied by Schweizer and Sklar.

1.8. Ingraham, 1969 :

Like Rosen, Ingraham [Ingraham 1967] proposed that every process of measurement of the coordinates of an event and more generally an observable field at that event by an observer $\mathcal{L}$ has an intrinsic lack of sharpness or dispersion, measured by fundamental length. This lack of sharpness or indeterminacy has nothing to do with the quantum mechanical indeterminacy. The later puts a lower bound on the errors of the simultaneously measured values of pairs of field components at an event; its mathematical expression is the non-commutativity of conjugate pairs of operators in Hilbert space. The lower bound is measured by Planck's constant. The proposed new indeterminacy is in addition to the quantum mechanical one. It would, for example, affects the measurement of even one of the conjugate pair of observable field components. This new indeterminacy would exist in classical as well as in quantum field theory. In fact, Pauli [Pauli 1923] pointed out the connection of the mathematical limitation of quantum electrodynamics with gravitation as mentioned by Landau and Klein [Landau 1965] i.e., $p \sim \frac{\hbar}{\ell_0}$ (p being the momentum with $\ell_0 = \sqrt{\chi \hbar c}$, where, χ is Einstein gravitational constant). It seems to hint at the indeterminacy in space-time of the light which is governed by probability laws in a quantized field theory, invariant with respect to the wider group of general relativity.

Now, if we adopt this idea of stochastic space-time as proposed by Ingraham then a singular Lorentz tensor in the old theory gives rise to a stochastic Lorentz tensor for each observer $\mathcal{L}$, but these are not in general the same tensor for observers $\mathcal{L}, \mathcal{L}', \mathcal{L}''$.... However, one can demand that equivalent observers $\mathcal{L}, \mathcal{L}', \mathcal{L}''$... can be able to communicate. The stochastic tensor representing the value of a certain physical field measured by $\mathcal{L}$ can be referred to coordinate system of $\mathcal{L}'$ and we can demand these operators or c numbers to be the result of $\mathcal{L}$'s measurement as reported to or as seen by $\mathcal{L}'$. The results of experiments performed by $\mathcal{L}$, can similarly be reported to any other observer $\mathcal{L}''$. We demand that $\mathcal{L}$ and $\mathcal{L}''$ both measured the same mean relative velocity of $\mathcal{L}$ with respect to $\mathcal{L}'$, so that the Lorentz transformations connecting them is unambiguous. More generally, the same mean velocity four vector of any macroscopic object is measured by all observers and all observers see light propagation with velocity equal to c. Ingraham discussed in details the Lorentz transformations among different stochastic observers and the way to calculate the mean values. He also pointed out the problems related to the four dimensional distribution functions in the stochastic space-time.

Rosen and Ingraham independently tried to understand the quantum field-theoretic structures and their divergencies. But both the programmes are far from complete. In both the cases, they introduced fundamental length to avoid the divergencies in field theory. It is interesting to note that if we look at the concept of random metric space (as introduced by Sherwood), i.e., given a E-space, for each ω in Ω, the function d_ω, defined on $S \times S$ by

$$d_\omega(p,q) = d(p(\omega), q(\omega)),$$

is a pseudometric on S but generally not a metric (since $p(\omega) = q(\omega) \to p = q$). It might help us to understand the above mentioned arbitrariness of the compositeness as well as the problems related to four-dimensional distribution. Almost during this period, Takano and Blokhintsev independently introduced the concepts of random metric spaces by considering the fluctuation in metric tensor. These approaches play significant roles in quantum field theories as well as in gravitational field.

1.9. New Developments, 1975 and Later :

In 1975, Blokhintsev [Blokhintsev 1975] started a systematic study of the stochastic spaces and published his results in a monograph. The causal relationship which was already there in the world of macroscopic bodies is carried over into the domain of microphysics. It gives rise to new insights into the foundational problems of geometry itself. In 1976, Carlton Frederick [Frederick 1976] published a profound paper on stochastic space-time and quantum theory. It has drawn a lot of attention regarding the applications of stochastic metric to explain the probabilistic nature of quantum theory. With a few assumptions, a large segment of the formalism of the quantum theory has been derived within this framework. According to his idea, the fluctuations of the space-time metric becomes less as the mass of the body becomes more. If at the position of the particle, the metric due to a specific physical situation is $g_{ik}^{(1)}$ and the metric due to a different physical situation is $g_{ik}^{(2)}$, then the metric due to the presence of both the physical situations is

$$g_{ik}^{(3)} = \frac{1}{2}\left(g_{ik}^{(1)} + g_{ik}^{(2)}\right) \tag{1.34}$$

This is known as the metric superposition principle. With the help of this superposition postulate, the interference phenomena in quantum mechanics can be described. In these cases, the actual points of space are of stochastic in nature. So the space considered by Frederick can be identified as a particular type of random metric spaces constructed by mathematicians. However, he did not discussed in details the stochastic metrics in four dimensional space-time so as to analyze the status of Lorentz transformations as well as the problems of divergencies in quantum field theory. Moreover, the effect of stochastic metric in Einstein's equation for gravitational field has not been investigated by Frederick. But it opens up new vistas in describing the behavior of microscopic particles.

Sisir Roy [Roy 1981] further developed this idea of stochastic space-time in the relativistic domain. He has tried to construct the relativistic equations for microparticles taking the fluctuation of space-time metric in cellular spaces. It helps us not only to understand the behavior of relativistic particles but also hints at solving problems related to the reduction of wave packet. It has had significant influence on the later developments of the problems of decoherence in quantum mechanics, as considered by Unturbe et al.[Unturbe 1992]. The semiclassical limit has also been studied within this framework. A kind of fuzzy set-theoretic notion has been introduced for understanding this limit problem. The stochastic cellular space is a special case of random metric space. This cellular structure has a striking similarity to that of Rosen's Quantum Geometry. The fluctuation of the metric of this cellular space helps us to understand the origin of the quantum potential [Roy 1986] which was not fully understandable in the Bohm-Vigier theory of hidden variables. During this period, mathematicians independently studied the connections between Fuzzy metric and Statistical metric spaces. Ivan Kramosil and Jiri Michalik [Kramosil 1975] investigated the notion of a fuzzy metric. This notion is then compared with that of statistical metric space, and both the conceptions are proved to be equivalent in certain sense. These results may help us to understand the extended structure of elementary particles and non-local fields within the framework of stochastic space-time in a more rigorous manner.

Vigier [Vigier 1982] tried to understand the structure of Dirac aether model of vacuum considering the fluctuation of space-time metric. It is very much helpful to understand the role of subquantum level in the framework of stochastic interpretation of quantum mechanics. Subsequently, Vigier and Roy [Vigier 1985] started a collaboration to understand the role of this fluctuating space-time model in quantum mechanics. Prugovecki [Prugovecki 1984] studied the stochasticity in the phase space so as to describe the quantum phenomena on a rigorous mathematical basis. Namsrai [Namsrai 1988] initiated his interesting investigation on stochastic quantum mechanics in stochastic metric spaces. Later he concentrated his efforts on studying the non local theory of quantized fields for scaler, electromagnetic and four-fermion interactions. His main emphasis was to show that the hypothesis of stochastic nature of space-time at small distances leads to the space-time metric non locality. The culmination of his efforts led to the publication of a beautiful monograph on *Non-local Quantum Field Theory and Stochastic Quantum Mechanics.* But most of these efforts were concentrated mainly in the domain of micro phenomena. Blokhintsev indicated that the density and pressure fluctuations of matter can produce fluctuation in the metric through Einstein equation for the gravitational field. In a series of papers, Nariai [Nariai 1975] investigated the mean value field of a highly turbulent gravito- hydrodynamic field (obeying Einstein equations) whose substratum is dominated by hadronic matter with large density fluctuations. In the case where the mean value metric represents the regular model universe, it is shown that the metric and the density fluctuation are compatible with his basic ideas. However, it is done by proposing a phenomenological modifications of the Einstein's equations.

In 1980, Gerald Rosen [Rosen 1980] studied Einstein equation considering the fluctuation of the conformal scalar field. The basic idea is that on the average, it gives rise to a Riemannian structure. This fluctuation of the metric tensor gives rise to the fluctuation of density and pressure through Einstein equation. It is also shown that the rate of expansion of the universe can be enhanced due to the fluctuation of the conformal metric. This approach of Rosen gives rise to new possibilities to study several important problems in Cosmology. Within this approach, Sinha et al. [Sinha 1987] have shown that the fluctuation of the scalar field gives rise to a non-zero and finite cosmological constant in Einstein equation. The recent experimental results from the Hubble space telescope [Freedman 1984] use a controversy regarding the age of the universe within the Big Bang cosmology and the importance of non-zero cosmological constant. The fluctuation of the space-time metric increases the rate of expansion in addition to the Hubble expansion as well as generates the non-zero cosmological constant. The above discrepancy can be re-examined within Rosen's framework. The conformal metric fluctuations have been subsequently considered by Bergia et al. [Bergia 1989] who claimed that:

(a) Ehrenfest's theorem of Quantum Mechanics is violated in as the much as expectation value of the geodesic displacement does not obey the classical equation; the kinetic energy of the test particle is not conserved.
(b) There is no way to obtain the inverse dependence on the mass of the test bodies of the diffusion coefficient, as a consequence of the fact that the equations of the theory of gravitation are at this stage independent of the mass of the test bodies.

But more and more voices are raised regarding the origin and universality of metric fluctuation. Some of the recent attempts [Roy 1992], directed towards these questions (which have a profound significance in classical as well as quantum mechanics) have been included and discussed in this book.

It is now clear from the various works at different decades that the fluctuation of space-time metric plays a significant role in understanding the behavior of microparticles and some aspects of astrophysics and cosmology. In this book, we have tried to understand the various physical phenomena within the framework of random cellular metric spaces. Here, the cells do not have sharp boundaries so that they can overlap each other. But the centers of the two cells can not approach each other indefinitely. A minimal distance exists between the centers of the cells. This length is the length scale of fluctuation and it is calculated as $10^{(-32)}$ cm by Blokhintsev et al. [Blokhintsev 1975] . This length is comparable to the size of the string like object. It raises lot of interesting new proposals to study different physical phenomena in nature. We have concentrated our discussion to the following areas :

(a) Statistical behavior of micro-particles and fluctuation of space-time.

(b) Random potential in classical and quantum mechanics.
(c) Astroparticle physics and Einstein's equation for gravitational field with fluctuating $g_{\mu\nu}$.
(d) Possible origin of fluctuation of Geometry.

STATISTICAL BEHAVIOR OF MICROPARTICLES

"The highest demonstration of reasoning that we have in any branch of knowledge can only make a fact probable, and nothing further. The most demonstrable facts of physical science are only probabilities - not facts yet. Facts are not only in the senses. Facts have to be perceived."

Swami Vivekananda : My Master.

2. STATISTICAL BEHAVIOR OF MICROPARTICLES

2.1. Microparticles and Quantum Mechanics :

At the beginning of this century, Einstein recognized how a physical phenomena like the Photo-electric effect may depend on individual quantum effects. On subsequent years he came to the conclusion that any radiation process involves the emission or absorption of individual light quanta or "photon" with energy and momentum

$$E = h\nu \quad \text{and} \quad p = h\sigma$$

respectively, where, h is Planck's constant, ν and σ are the number of vibrations per unit time and the number of waves per unit of length respectively. Now, the idea of photon raises an unforeseen dilemma since corpuscular picture of radiation would be irreconcilable with interference effects. The interference phenomena is an essential part of radiative phenomena which can be described only in terms of a wave picture.

This situation can be understood only by using the concept of probabilities i.e., to estimate probabilities for the occurrence of individual radiation processes. However, this probability law is quite different from the familiar application of statistical considerations as practical means of accounting for the properties of mechanical system of great structural complexity. Bohr pointed out : "In Quantum physics we are presented not with intricacies of this kind, but with inability of the classical frame of concept to comprise the peculiar feature of indivisibility or characterizing the elementary processes". The failure of theories of classical physics was further developed by the progress of knowledge of atomic structure. In fact, Planck himself in his treatment of thermal radiation essentially introduced the statistical character but he was refrained from definite conclusions as to the extent to which the existence of the quantum implied a departure from the foundations of mechanics and electrodynamics.

During the first meeting of Einstein with Niels Bohr in 1920, the corpuscular and wave nature of quantum involving radiation processes were the main theme of their conversation. Subsequent discoveries in atomic physics made it clear that no simple picture of corpuscular collision could offer an exhaustive description of the radiation processes. The fundamental discovery of de Broglie in 1925 paved the way to the development of a more comprehensive quantum theory. De Broglie recognized that wave-particle duality is not confined to radiation but is equally unavoidable in describing the behavior of material particle which subsequently was confirmed by experiment. The decisive steps by Schrödinger on wave equation and Heisenberg matrix mechanics in formulating uncertainty relation made it clear that the theories of classical physics were untenable to describe the behavior of microparticles. The crucial point of the main theme of the then discussion was the impossibility of any sharp separation between the behavior of the atomic object and the interaction with the measuring instruments which serve to define the conditions under which the phenomena appear.

The introduction of the complementary principle helped Bohr to classify the situation in a major way. Now wave and particle aspects may be considered as the complementary aspect of the same object. Since the inception of this modern quantum theory, a great debate was going on regarding the statistical interpretation of quantum mechanics. In the spirit of Einstein, Bohm, de Broglie and his collaborators tried to describe this behavior starting with Hamilton-Jacobi equation and were able to derive Schrödinger wave equation. This is popularly known as Hidden variable theory. Afterwards, different variants of hidden variable theory were also propounded by various authors. But it should be emphasized that in his original formulation of hidden variable theory, Bohm himself considered the interaction of the microparticles with the hidden random bath. This may be the origin of "quantum force or potential" which is crucial in their framework. The concept of the hidden bath which gives rise to the quantum force is again irreconceivable in classical physics. So, even within the framework of hidden variable theory, the classical statistical consideration is unable to accommodate the new phenomena arising out of experimental discoveries around 1920.

In 1966, Edward Nelson proposed the hypothesis of background field in deriving Schrödinger equation within the framework of classical probability theory. According to this hypothesis, the background field causes a physical system to undergo a diffusion process with diffusion constant $D = \frac{\hbar}{2m}$ satisfying a variational principle

$$\delta E(\int L dt) = 0 \tag{2.1}$$

E denoting the expectation value.

Though the prediction of Nelson's [Nelson 1967] framework agree with that of usual quantum mechanics, yet it attempts to provide a realistic , objective description of physical events in classical terms. For example, no physical system of finitely many degrees of freedom is truly isolated, it is always in interaction with background field. As a result, a number of problems that were solved by Quantum Mechanics or at least were declared non-problems, are reopened.

The probability theoretical approach to non-relativistic quantum mechanics is introduced by Nelson, subsequently developed by Guerra and Ruggiero [Guerra 1973] to deepen our understanding of space-time with microlocal structure, the microscopic domain of less than 10^{-17} cm in length. Several attempts [Caubet 1975]; [Aurbach 1971]; [Yasue 1977]; [Vigier 1979]; [Lehr 1977]; [Roy 1979, 1980] have been made to extend Nelson's procedure in the relativistic domain. One of the main difficulties is that the diffusion coefficient ($\hbar/2m$) vanishes at relativistic limit as pointed out by Hakim [Hakim 1968]. Moreover, Nelson proposed the existence of a random background field but he did not consider its inclusion explicitly in the formalism. This problem is present in both non-relativistic as well as in relativistic region.

Several authors tried to introduce the idea of background field in Nelson's mechanics by introducing the fluctuating space-time metric. Frederick [Frederick 1976] started a systematic investigation using the fluctuation of space-time metric to study the different aspects of non-relativistic quantum mechanics. His idea was further developed by other authors [Roy 1981]; [Namsrai 1986] and generalized to the relativistic domain. Bergia et.al. [Bergia 1989] discussed the problems and perspectives of stochastic mechanics and the role of background field in a recent review.

2.2. Fluctuation of Space-Time Metric :

In chapter **1**, we have already discussed the basic idea of stochastic geometry. The basic assumption of stochastic geometry is that, to each pair of elements A, B of stochastic space $R(A)$, it is possible to associate a distribution function $F_{AB}(z)$ which can be interpreted as the probability that the distance between the points is less than z. In the domain of physics, it is possible to consider at least four cases which lead to stochastic spaces :

(a) The ideal instrument for measuring distances has an irremovable indeterminacy say $\Delta x \approx l_0$

(b) The probes i.e., the particles by which the points in space are fixed, have finite extension l_0 and this may lead to $\Delta x \approx l_0$.

(c) The vacuum itself is a kind of turbulent medium so that the signal velocity fluctuates.

(d) In those regions of space, where, there are turbulent motion of matter accompanied by fluctuating density of matter or very irregular velocities of motion, may be also characterized by the fluctuating metric.

In the last two cases i.e., in **(c)** and **(d)**, the medium is considered as turbulent. But in one case due to the fluctuation of the vacuum, the matter may be in irregular motion and the reverse in other one. Let us first discuss the fluctuation of the space-time metric in non-relativistic domain. In such case, the space-time is considered as stochastic in absence of matter. The more mass in the space-time, the less stochastic the space-time becomes. In the absence of matter, the space-time is stochastic but not necessarily flat. Following Frederick [Frederick 1976], let us introduce the following postulates :

1. Metric Probability Postulate :

$$P(x,t) = A\sqrt{-g} \tag{2.2}$$

where, for a one particle system, $P(x,t)$ is the particle probability distribution. A is a real valued function and g is the determinant of the metric.

2. Metric Superposition Postulate :

Let $g_{ij}(1)$ and $g_{ij}(2)$ denote the metrics for a particle at position 1 and position 2 under different physical conditions.Then a metric $g_{ij}(3)$ can be written as

$$g_{ij}(3) = \frac{1}{2}[g_{ij}(1) + g_{ij}(2)] \tag{2.3}$$

3. Metric Ψ - Postulate :

There exists a local complex diagonal coordinate system in which a component of the metric can be associated to the location of the particle in the wave function Ψ. For simplicity, consider the distribution of each metric component at any point θ be normal

$$f_{\bar{g}_{ij}}(g_{ij}) = \frac{1}{\sigma(2\pi)^{1/2}} exp[-1/2(\frac{g_{ij}}{\sigma})^2] \tag{2.4}$$

Within this framework, it is possible to establish the following uncertainty principle for position and momentum observables q and p respectively.

$$\Delta q^1 \Delta(p_{\nu\min} g^{\nu 1}) = \Delta q^1 \Delta p^1 > k_{\min} \tag{2.5}$$

In quantum mechanics, the wave function contains more information rather than the probability density. The wave function contains the phase which is very important in describing the interference phenomena. But here, in the frame of stochastic space-time we are dealing directly with $P(x,t)$. So, it appears to be problematic to explain the interference phenomena within this framework. Let us consider a two slit experiment in stochastic space-time. According to metric superposition postulate, the metric ${g_{ij}}^{S3}$ can be written as

$${g_{ij}}^{S3} = \frac{1}{2}({g_{ij}}^{S1} + {g_{ij}}^{S2}) \tag{2.6}$$

${g_{ij}}^{S1}$ and ${g_{ij}}^{S2}$ being the metric tensors when the slit one is open only and slit two open only respectively. It is also assumed that the screen in the experiment is placed far from the slits so that the individual probabilities $(-g^{S1})^{1/2}$ and $(-g^{S2})^{1/2}$ can be considered as constants over the screen.

Now, if the particle is propagated in , say, the x^3-direction and, of course, also in the x^4-direction, we might expect that the metric be equal to the Minkowski metric $\eta_{\mu\nu}$. We can think a situation where the metrics ${g^{S1}}_{\mu\nu}$ and ${g^{S2}}_{\mu\nu}$ are in the average $\eta_{\mu\nu}$.

Consider the following average metric

$${\bar{g}^{S1}}_{\mu\nu} = \left\| \begin{matrix} 1 & 0 & 0 & 0 \\ 0 & 1 & 0 & 0 \\ 0 & 0 & s & 0 \\ 0 & 0 & 0 & -t \end{matrix} \right\|$$

and

$$|{\bar{g}^{S1}}_{\mu\nu}| = -st \tag{2.7}$$

s and t are yet unspecified function of position.

If $s = t^{-1}$,then,

$$|\overline{g}^{S1}{}_{\mu\nu}| = |\eta_{\mu\nu}| = -1 \qquad (2.8)$$

Again taking $s = \exp(i\alpha)$, where α is some unspecified function, we get

$$\overline{g}^{S1}_{\mu\nu} = \begin{Vmatrix} 1 & 0 & 0 & 0 \\ 0 & 1 & 0 & 0 \\ 0 & 0 & e^{i\alpha} & 0 \\ 0 & 0 & 0 & -e^{-i\alpha} \end{Vmatrix}$$

and,

$$\overline{g}^{S2}{}_{\mu\nu} = \begin{Vmatrix} 1 & 0 & 0 & 0 \\ 0 & 1 & 0 & 0 \\ 0 & 0 & e^{i\beta} & \\ 0 & 0 & 0 & -e^{-i\beta} \end{Vmatrix}$$

where, β is another unspecified function.

Then after simple calculation we get

$$(-|\overline{g}^{S3}_{\mu\nu}|)^{1/2} = \frac{1}{2}|\cos\frac{1}{2}(\alpha - \beta)| \qquad (2.9)$$

$|,|$ denotes the absolute value.

This is nothing but the phenomenon of interference. But the use of complex functions in the metric is unphysical as the resultant line element

$$ds^2 = \overline{g}_{\mu\nu} dx^\mu dx^\nu \qquad (2.10)$$

would be complex and unphysical. Now one can raise a question : **Is it possible to reproduce the previous arguments but with real functions ?** The answer is **yes.**

Let

$$\begin{Vmatrix} dx^1 \\ dx^2 \\ dx^3 \\ dx^4 \end{Vmatrix} \qquad (2.11)$$

and

$$G = ||\overline{g}_{\mu\nu}||$$

Then

$$X^tGX = ds^2 = \bar{g}_{\mu\nu}dx^\mu dx^\nu \tag{2.12}$$

where X^t is the transpose of X. We shall consider the transformations which will leave the line element ds^2 invariant. Given a transformation matrix W,

$$X = WX'$$

and

$$X^tGX = X'^tG'X' = (X^t(W^t)^{-1})G'(W^{-1}X)$$

However, $X^tGX = X^t(W^t)^{-1}(W^tGW)(W^{-1}X)$ so that $G' = W^tGW$. In the transformed coordinates a matrix $\bar{g}^{S1}_{\mu\nu} = G^{S1}$ goes to $W^tG^{S1}W$.

Therefore,

$$\Psi_1^*\Psi_1 = (-\mid W^tG^{S1}W \mid)^{\frac{1}{2}} = (-\mid W^t \mid\mid \bar{g}^{S1}_{\mu\nu} \mid\mid W)^{\frac{1}{2}} \tag{2.13}$$

$$\Psi_3^*\Psi_3 = (-\frac{1}{16} \mid W^t \mid\mid \bar{g}^{S1}_{\mu\nu} \mid\mid W)^{\frac{1}{2}} = (-\frac{1}{16} \mid W^t \mid\mid G^{S1} + G^{S2} \mid\mid W \mid)^{\frac{1}{2}} \tag{2.14}$$

Now if we can find a transformation matrix W with the properties

- **(a)** $\mid W \mid = 1$.
- **(b)** W is not a function of α or β
- **(c)** W^tGW is a matrix with only real components, then we will again have the interference phenomenon with $\bar{g}'_{\mu\nu}$ real.

Again

$$\Psi_1^*\Psi_1 = \Psi_2^*\Psi_2 = 1; \quad \text{and} \quad \Psi_3^*\Psi_3 = \frac{1}{2}\left|\cos\frac{(\alpha-\beta)}{2}\right|$$

The appropriate matrix W is

$$\left\|\begin{matrix} 1 & 0 & 0 & 0 \\ 0 & 1 & 0 & 0 \\ 0 & 0 & \frac{i}{\sqrt{2}} & \frac{1}{\sqrt{2}} \\ 0 & 0 & \frac{1}{\sqrt{2}} & -\frac{i}{\sqrt{2}} \end{matrix}\right\|$$

As previously, if

$$\left\|\begin{matrix} 1 & 0 & 0 & 0 \\ 0 & 1 & 0 & 0 \\ 0 & 0 & e^{i\alpha} & 0 \\ 0 & 0 & 0 & e^{-i\alpha} \end{matrix}\right\| = \left\|\bar{g}^{S1}_{\mu\nu}\right\|$$

Then

$$\left\| \begin{matrix} 1 & 0 & 0 & 0 \\ 0 & 1 & 0 & 0 \\ 0 & 0 & -\cos\alpha & \sin\alpha \\ 0 & 0 & \sin\alpha & \cos\alpha \end{matrix} \right\| = \left\| \bar{g}^{S1}_{\mu\nu} \right\| = W^{+}\bar{g}^{S1}_{\mu\nu}W$$

so that in order to reproduce the phenomenon of interference, the stochastic metric $\bar{g}_{\mu\nu}$ will have off diagonal terms. The coordinates appropriate to $\bar{g}^{S1}_{\mu\nu}$ are

$$\begin{aligned} x^{1'} &= x^1 \\ x^{2'} &= x^2 \\ x^{3'} &= -\frac{i}{\sqrt{2}}x^3 + \frac{1}{\sqrt{2}}x^4 \\ x^{4'} &= \frac{1}{\sqrt{2}}x^3 - \frac{i}{\sqrt{2}}x^4 \end{aligned} \tag{2.15}$$

So by choosing appropriate coordinate system which is complex, we can treat the free space probability distributions $\Psi^*\Psi$ in a simple way. But it raises number of difficult problems. For example, the stochastic metric g_{ij} can not be generalized to real space-time simply by increasing the number of dimensions. The main difference between the physical theory of stochastic space-time and the mathematical construction as discussed in chapter 1 is that the physical space has an indefinite metric where as, the metric described by the distributions $F_{AB}(z)$ is assumed to be positive definite. In physical stochastic space-time, frame of references defined in stochastic space-time should be connected by Lorentz transformations. This indicates that the four dimensional probability for stochastic metric $g_{\mu\nu}$ should be function of an invariant like

$$(T-t)^2 - (X-x)^2 \tag{2.16}$$

However, in that case, the probability is not normalizable and can not represent the distribution of stochastic variables x. Blokhintsev [Blokhintsev 1973] constructed another invariant by which the probability can be written as a function of this invariant. Then the problems like normalizability etc. can be avoided. He proposed the idea of a time-like vector n as well as the four-dimensional vector $(X-x)$. Then it is possible to write a positive-definite distance R as

$$R = +\sqrt{2\{n,(X-x)\}^2 - s^2} \geq 0 \tag{2.17}$$

But, the choice of this time-like vector is another problem. Blokhintsev himself was aware of this problem and he discussed the two possibilities to choose this time-like vector :

(a) The n is tied to the reference frame of the observer so that the direction of n corresponds to the direction of the world line of the observer. The n has components $n = (1,0,0,0)$ for all observers. Therefore the probability density will not be invariant under changes in coordinate systems. So special care must be taken to quantities where invariance is not in doubt; for example, the total probability of forming a specific number and type of particles. This quantity may depend on the reference frame only in the case where a special reference frame exists. This is a kind of vacuum which acts as a medium, like "aether" in the old sense.

(b) If the particles are taken as extended in space-time instead of point, then stochastic nature of the coordinates may depend also on this particle dimension. The coordinates now become stochastic because it is not possible to define which point inside the particle scattered the signal, as if, it were not localized accurately in the space-time. Here, particles can be considered as hazy lumps instead of points. This is similar to the idea of C-space [Schweizer 1983] as already developed by mathematicians. As a special case, this n can be identified with a four-dimensional particle velocity u. Now relativistic invariance will be not a problem as the velocity u satisfies the Lorentz transformation. In general, we may think of this extra variable as internal variable. Here lies the deeper connection of non-local fields and the stochastic space-time which we shall discuss in latter chapter.

The assumption that the variables u and $\xi = X - x$ commute,

$$[u, \xi] = 0 \tag{2.18}$$

is implicit in the assumption that the probability depends on these variables. Therefore, it is natural to consider the variable ξ as the internal variable. A stochastic space-time then be constructed with the line element defined by

$$dx^2 = G_{\mu\nu}(x, \xi) dx^\mu dx^\nu \tag{2.19}$$

Here, the metric tensor $G_{\mu\nu}$ is function of both x and ξ. This is popularly known as Finsler metric [Asanov 1985]. We shall show in latter chapter that Finsler metric is inherently probabilistic in nature. If we reformulate Nelson's Mechanics [Roy 1979] in 4-Dimensional Finsler Manifold, it immediately solves several problems which are present in relativistic domain. The stochastic process is described in 4-Dimensional cellular domain characterized by Finsler metric. It has been shown [Schewizer 1983] that there exists a minimal distance between the centers of two cells of hazy lumps.

This finite length scale makes the diffusion coefficient as finite and non-zero even in the relativistic domain. These lumps may be thought of as seat of elementary particles or string like objects. Now it is possible to show that we can get correct relativistic wave equations if stochastic processes are defined in a multiply connected space-time where quantum of action is one dimensional period of one form ω_2 i.e.,

$$\oint \omega_2 = n\hbar \tag{2.20}$$

analogous to Bohr-Sommerfeld quantization schemes. it is shown that there exists a function

$$S : F \longrightarrow R \tag{2.21}$$

such that $\omega_2 = d(\hbar S)$, F being the generalized four dimensional differentiable Manifold where the stochastic process is defined and R is real line. It is well known that according to Bohr-Sommerfeld quantization, one form

$$\omega = pdq - Hdt$$

is not exact i.e., there exists some topological constraints such that it can not be shrunk to a point. So the space-time may be thought of as a multiply connected region of stochastic fields with period $\hbar$ embedded in a simply connected space-time. In this picture, space-time continuum is considered as a set of hazy lumps or C-spaces. Each hazy lumps or "cloud" is associated with a distribution function and thus a concept of probabilistic distance emerges. One may consider some kind of random motion in the cellular space-time so as to measure only the mean path which relates to the center of gravity of the hazy lump. The mean velocity can be evaluated generally of magnitude smaller than c, the speed of light, but the deviation from the mean path can be regarded as a sort of Brownian motion propagating with the speed of light. This splitting of motion into a systematic part and a fluctuation is similar to what is done in Langevin equation with difference that the fluctuation can no longer be taken to be independent of mean velocity because of the constraint imposed by fixed c. The introduction of the internal variable ξ in the metric is physically equivalent to anisotropic space. This internal variable can be linked up to the anisotropy of space-time. Let us designate this variable as ν.

In general, the wave function satisfying relativistic wave equation can be written as

$$\Psi(q,\nu,\tau) = exp[R(q,\nu,\tau) - iS(q,\nu,\tau)] \tag{2.22}$$

and

$$\Psi^*(q,\nu,\tau)\Psi(q,\nu,\tau) = \rho(q,\tau) \tag{2.23}$$

$q, \nu \in F$, F being four-dimensional Finsler space and τ denoting the proper time.

The probability density, here, ρ is assumed to be independent of ν, because otherwise, we will not get the probability density for Ψ function in quantum mechanics. It is shown that if the probability density depends on ν, then the gauge group $U(1)$ of quantum mechanics has to be replaced by higher order group. This framework give rise to a new insight in understanding quasi-classical limit from the geometrical point of view.

There are two kinds of [Skawianowski 1977] asymptotics describing the correspondence principle which is as follows :

- **Large quantum numbers**
- **Small value Planck constant.**

They are supposed to be essentially equivalent, however, up to now there is no rigorous and general proof of this equivalence. Besides, in spite of some current views, neither of these methods leads automatically to classical laws when starting with quantum theory. Some kind of physical intution and feeling is necessary to avoid mistakes; there are some dangers and traps typical in either of these descriptions.

In a crude way, the asymptotics of large quantum numbers consists in the limit transition

$$n_0/\Delta n \rightarrow \infty \tag{2.24}$$

where, n_0 is a mean quantum number of a physical system and Δn is spread of quantum numbers. Technically speaking;

$$n_0 >> \Delta n >> 1 \tag{2.25}$$

Quantum formulae should approach then those derived from the classical laws. Technically, the $h \rightarrow 0$ approach is much easier and more "automatic" than that based on $n \rightarrow \infty$. In fact, the transition $h \rightarrow 0$ transforms formally the whole conceptual structure of the quantum theory into a classical one, but it has nothing to do with the real laboratory conditions of "quasi-classicality". Physics is interested rather in answering the questions :

What are quasi-classical situations in the real world, when the Planck constant has a fixed value ?

It is often argued that it is only the $n \rightarrow \infty$ asymptotics which are able to answer this question. But we believe that this is not true in general. Consider Ψ_n as the wave functions of stationary states of a bounded system. Here, n can take only a discrete set of values. But in case of classical object, n takes continuous arbitrary values, thus even in case of large n there remains the same old problem of real laboratory conditions of quasi-classicality as with $h \to 0$ asymptotics. Moreover, it is not true for large values of n that Ψ_n becomes quasi-classical through out the whole configuration space.

With this situation in mind let us consider our scheme of relativistic stochastic quantization, where $\hbar$ is considered as the non-vanishing period of one form ω_2 i.e.,

$$\oint \omega_2 = n\hbar \tag{2.26}$$

Here n can take only integer values. Now, whenever we approach the classical limit, in the laboratory we usually use cluster of atoms and molecules to form a massive object. Again, as we attach more and more atoms and molecules, the well defined boundaries defined on the multiply connected space-time become more and more fuzzy. Technically speaking, the topology becomes more and more probabilistic rather than the usual topology. In mathematical language this can be expressed by saying that integral cycles upon which the form ω_2 is defined should be extended to the real cycles.

It is well known in the integration theory that if we consider only chains $c = \sum a_i \sigma_i$ which are sums of simplices with integer coefficients on the manifold M of any dimension, then we talk about integer chains, integer cycles and integer boundaries. According to De Rham's theorem, the integer periods

$$\int_z^\omega$$

of a closed form ω are the periods taken over integer cycles only. Now we can introduce the notion of a real cycle in the following way :

If $\Gamma_p'^1 \cdots, \Gamma_p'k$ are any k integral p-cycles, the object $\Gamma_p' = \sum_1{}^k \lambda_i \Gamma_p{}'^i$, where $\lambda_1 \cdots, \lambda_k$ are real numbers, is called a real p-cycle and we define

$$\int_{\Gamma'} \omega = \sum_i \lambda_i \int_{\Gamma'^i{}_p} \omega \tag{2.27}$$

Now by considering the one form ω_2 defined over the real cycle, it can be shown that the period will be represented by arbitrary real numbers. As a result, the characteristic number n (as radial quantum number n_r in case of Bohr-Sommerfeld quantization $\oint p_r = n_r \hbar$) can be arbitrary real numbers, i.e., the so called quantum levels will be closely spaced. To a certain extent when the degree of fuzziness is maximum, the period will vanish i.e. ω_2 can be defined globally. As we come at the macrolevel from the microstage, the holes in the space-time to be blurred i.e. the discrete space-time become to be more and more continuous. So, the integral cycles in the microstructure will be more and more broken into non-integral cycles and the boundaries of the holes will be more and more fuzzy. Finally, at the macrolevel where ω_2 can be written globally, space-time seems to be simply connected. Of course, in the strict mathematical sense, one can use the concept of fuzzy sets and fuzzy topology to describe the above picture in a more rigorous fashion.

Now one can look at the formal similarity between the $n \to \infty$ asymptotics and the maximum degree of fuzziness. $n \to \infty$ implies higher and higher modes i.e., the increasing degree of fuzziness of the topology so that in the limit ω_2 can be written globally and we have the corresponding Hamilton-Jacobi equation for

$$\Psi = \sqrt{\rho} \exp{(i/\hbar)S]} \tag{2.28}$$

where $\rho = \exp[2R]$.

It should be mentioned that Kramosil and Michalek [Kramosil 1975] have already established that the concepts of statistical metric and fuzzy metric are equivalent in certain sense. So the above procedure can be thoroughly considered in fuzzy metric spaces which might have greater impact in other branches of physical science.

It is worthmentioning that recent developments on quantum dynamics with two Planck constants [L.Diosi, 1995] shed new light on semiclassical dynamics. The basic motivation of this approach is as follows : Usually quantum dynamics is thought to be as fundamental one and classical dynamics is considered as a special limit of it. But there exists at least one classical dynamics whose quantization is still problematic. Gravity is an example of such a system. No consistent quantization framework exists so far in case of gravity. We can take it as classical so that the classical dynamics of graviatation would couple with quantum dynamics of other fileds. The mathematical difficulties are not trivial. Now the problem is whether classical and quantum dynamics could be coupled at all in mathematically consistent scheme. For this let us start with a quantum system composed of two canonical subsystem and assume two different Planck constants for each subsystems so that

$$\begin{aligned} [q_1, p_1] &= i\hbar_1 \\ [q_2, p_2] &= i\hbar_2 \end{aligned} \tag{2.29}$$

with, say, $\hbar_1 > \hbar_2$.

At the end, we shall take the limit $\hbar_1 \to \hbar$ and $\hbar_2 \to 0$. In such a way we shall obtain a (hybrid) semiclassical dynamics where (q_1, p_1) are quantum and (q_2, p_2) are classical. However, it does not lead to any conflict as long as the states ρ_n of two systems $(n = 1, 2)$ evolve independently of each other according to their Schrödinger equations

$$\dot{\rho}_n = -(\frac{i}{\hbar_n})[H_n, \rho_n] \tag{2.30}$$

where $H_n = H_n(q_n, p_n)$ are corresponding Hamiltonian. Now the total Hamiltonian can be written as

$$H = H_1 + H_2 + H_I \tag{2.31}$$

where the interaction Hamiltonian H_1 can be expanded into a series of interacting currents

$$H_1(q_1, p_1; q_2, p_2) = \sum_\alpha J_1^\alpha(q_1, p_1) J_2^\alpha(q_2, p_2); \qquad (J_i^\alpha, J_2^\alpha \neq 1) \tag{2.32}$$

In fact. all dynamical variables of the composed system can be decomposed in a similar form :

$$A(q_1, p_1; q_2, p_2) = \sum_\alpha A_1^\alpha(q_1, p_1) A_2^\alpha(q_2, p_2) \tag{2.33}$$

A suitable generalizatioin of the quantum bracket for the composed system with two Planck constants can be made with ansatz

$$\{A, B\}_Q \equiv -\frac{i}{2\hbar_1}[A_2, B_2] + [A_1 + B_1] - \frac{i}{2\hbar_2}[A_1, B_1] + [A_2, B_2] \tag{2.34}$$

Then the Schrödinger equation of motion can be written as follows :

$$\dot{\rho} = -\frac{i}{\hbar_1}[H_1, \rho] - \frac{i}{\hbar_2}[H_2, \rho] - \frac{i}{\hbar av}[H_1, \rho] + \frac{i}{2}\Delta\hbar^{-1}(J_1\rho J_2 - J_2\rho J_1) \tag{2.35}$$

where $\hbar_{av} = 2\hbar_1\hbar_2/(\hbar_1 + \hbar_2)$ and $\Delta\hbar^{-1} = \hbar_2^{-1} - \hbar_1^{-1}$.

The first three terms on right hand side of eqn.(2.35) generate unitary evolution. But the fourth term distorts this unitary evolution. However, it would not create serious problem if it does not violate positivity of ρ.

But it does. In this situation we should not ask too sharp questions in the presence of dynamics $\hbar_1 \neq \hbar_2$. We can enquire about blurred values of the dynamical variables. As a result, we need certain smoothing procedure. Diosi [Diosi, 1995] considered a smoothing mechanism by hand. The interaction Hamiltonian is replaced by a noisy one

$$H_1^{\text{noise}}(q_1, p_1; q_2, p_2) = (J_1(q_1, p_1) + \delta J_1(t))(J_2(q_2, p_2) + \delta J_2(t)) \quad (2.36)$$

where $\delta J_1, \delta J_2$ are classical noises superposed on the interacting currents J_1, J_2. Their correlations may be assumed as

$$< \delta J_n(t')\delta J_n(t) >_{\text{noise}} = \frac{1}{2}\hbar_m^2 \Delta\hbar^{-1}\lambda_n \delta(t' - t) \quad (2.37)$$

with $(n, m) = (1, 2), (2, 1)$.

with $\lambda_1\lambda_2 = 1$. The two noises $\delta J_1, \delta J_2$ are assumed to be independent of each other. The total Hamiltonian becomes noisy as

$$H^{\text{noise}} = H_1 + H_2 + H_I^{\text{noise}} \quad (2.38)$$

The blurred dynamics is defined by the Schrödinger quantum of motion averaged over the noise:

$$\dot{\rho} = < \{H^{\text{noise}}, \rho\} >_Q$$

After some calculations, we obtain the master equation as

$$\begin{aligned}\dot{\rho} &= -\frac{i}{\hbar_1}[H_1, \rho] - \frac{i}{\hbar_2}[H_2, \rho] - \frac{i}{\hbar_{\text{av}}}[H_1, \rho] \\ &- \frac{1}{2}\Delta\hbar^{-1}(iJ_2\rho J_1 - iJ_1\rho J_2) \\ &+ \frac{1}{2}\lambda_2[J_1, [J_1, \rho]] + \frac{1}{2}\lambda_1[J_2[J_2, \rho]) \end{aligned} \quad (2.39)$$

This master equation is utilized to construct semiclassical dynamics. As per our plan, to make the sysytem (q_2, p_2) classical we need the limit $\hbar \to 0$ with zero spread wave packets $| \, q_2, p_2 >$.

We also assume that the density operator ρ is diagonal in this basis

$$\rho = \int \rho(q_2, p_2) \mid q_2, p_2 >< q_2, p_2 \mid dq_2 dp_2 \quad (2.40)$$

Let us rescale the λ-coefficient :

$$\lambda_2 = \lambda\hbar_2, \quad \lambda_1 = \frac{\lambda^{-1}}{\hbar_2}$$

Performing the limit $\hbar_2 \to 0$ on the diagonal element $< q_2, p_2 \mid \cdots \mid q_2, p_2 >$, the resulting semiclassical master equation takes the form

$$\begin{aligned} \dot{\rho} &= -\frac{i}{\hbar}[H,\rho] + \frac{1}{2}\{H,\rho\}_P - \frac{1}{2}\{\rho, H\}_P \\ &- \frac{1}{4}\lambda[J_1,[J_1\rho]] + \frac{1}{4}\lambda^{-1}\{J_2,\{J_2,\rho\}_P\}_P \end{aligned} \tag{2.41}$$

where, in the total Hamiltonian $H(q_1, p_1, q_2, p_2)$, the variables q_1, p_1 are operators while the variables q_2, p_2 are numbers. Obviously, J_1 is operator and J_2 is number.

It is now clear from the above master equation that it is the density operator of the quantum subsystem and phase space distribution of the classical subsystem. Its trace over the quantum subsystem's states yields the phase space distribution $\rho_2(q_2, p_2)$ of the classical subsystem while its integral over q_2, p_2 yields the reduced density operator ρ_1 of the quantum subsystem. This semiclassical master equation can be applied to study the interaction of quantized nonrelativistic matter with weak classical gravitational field. It should be noted that the coupled classical quantum dynamics is irreversible in contrast to the pure classical or pure quantum canonical dynamics. This approach can be applied to interacting quantum systems whose particular subsystems behave classically under certain conditions. However, in each case, the mechanism of irreversibility is not unique. This framework gives rise new insights in understanding the semiclassical dynamics but more works are needed to clarify the situation by considering the systems in relativistic regime.

2.3. Decoherence in Quantum Mechanics :

It is well known that the interaction of the system subjected to observation with an apparatus alone does not, constitute a measurement, since it must be supplemented by the act of looking to see what the outcome of the interaction has been or by some automatic record of the result. It is usually the sum of two requirements :

- That there shall be an interaction with a measuring device.
- That a number, observed or recorded, shall result from the interaction.

The word "number" is, of course, taken in a nonliteral sense, yes and no, heads and tails, also many other characteristics of direct observations are numbers as they are translable into numbers. But the crucial problem is that whether quantum mechanics can describe the measurement process which in principle belongs to the domain of classical physics. Usually there are two ways to solve the measurement problem of quantum mechanics. Stapp [Stapp 1992] described these as "Heisenberg" and "Von Neumann" reductions in the respective cases that a particular result is picked up from all outcomes potentially present in the wave function ("Heisenberg") or where the coherence between off diagonal elements of the density matrix, in some suitable representation, is destroyed ("Von Neumann").

In the "conventional" or "Copenhagen" interpretation, either the projection postulate or the existence of a suitable stochastic mechanism that operates beyond quantum mechanics (and then responsible for quantum jump) are proposed. On the otherhand, in "Von Neumann" approach, one needs not go beyond quantum mechanics as far as statistical (i.e., "ensemble") interpretation is concerned.

Heisenberg's idea has recently been developed in a mathematical form by Ghirardi, Rimini and Weber [Ghirardi 1986].

Ghirardi et al. [Ghirardi 1986] first proposed a non-Hamiltonian dynamics for the description of macroscopic objects and then studied whether this can be obtained by modifying the standard quantum dynamics for their microscopic constituents. If such a modification leaves practically unaltered the behavior of microscopic systems as accounted for by quantum mechanics, a possible unified description emerges to account for both quantum and classical behavior of microscopic and macroscopic systems. It is assumed that localization process occurs individually for each constituent of a many particle system. Let us consider a system of N particles in one dimension. Denoting by λ_i the frequency of the process for the constituent i and assuming that the accuracy of the localization is the same for all constituents, the evolution equation for the composite system is given by

$$\frac{d}{dt}\rho = -\frac{i}{\hbar}[\hat{H}, \rho] - \sum_{i=1}^{N} \lambda_i(\rho - T_i[\rho]) \tag{2.42}$$

where

$$T_i[\rho] = \sqrt{(\alpha/\pi)} \int_{-\infty}^{\infty} dx e^{-(\alpha/2)(\hat{q}_i - x)^2} \rho e^{-(\alpha/2)(\hat{q}_i - x)^2}$$

and $\hat{q}_i$ is the position operator for the i-th constituent of the system.

Now studying the above equation it is possible to have a natural solution to the puzzling situation originating from the occurance of linear superpositions of differently located states. But the physical origin of these localizations is not considered within the approach. In this sense it is known as spontaneous localization procedure. They proposed that the coordinate-space wave function of each particle in the universe has a certain innate chance to spontaneously collapse to a new form. This "Spontaneous stochastic localization" mechanism of GRW brings the physical character of the wave function at the macroscopic level in accord with our sense impressions. Stapp [Stapp 1992] has rightly pointed out that

"the mechanism is ad hoc : it is not connected to anything else in physics".

Pearle [Pearle,1989] made an attempt to improve this situation. In his proposal, a quantum system is coupled weakly to a "white noise" function $W_{\Omega}(x,t)$, which has the property

$$<< W_{\Omega}(x',t')W_{\Omega}(x'',t'') >>= \phi(x'-x'')\delta(t'-t'') \tag{2.43}$$

Here,

$$\phi(x'-x'') = exp[-(x'-x'')^2/2\sigma] \tag{2.44}$$

is the Gaussian function of *GRW*. Symbol Ω designates a particular white noise function $W_{\Omega}(x,w)$, which is associated with a probability function $P(\Omega)$, and $<<>>$ represents the weighted sum over all Ω. $\delta(t'-t'')$ is the characteristic of white noise. The function $\phi(x'-x'')$ characterizes the correlation of the noise at neighboring points. This model is less ad hoc than *GRW* as noise is present in any physical system. Stapp [Stapp 1992] gave a simpler derivation of Pearle's results. He considered an interesting case i.e., the noise associated with 2.7°K background radiation and the reduction of the wave packet due to this noise under certain circumstances. This model appears to be more realistic from the point of view of a physical system and its evolution. But it is still far from the exact laboratory conditions under which the reduction happens.

Very recently, Rauch et al.[Rauch 1995, 1996] studied the vanishing of interference in phase space using the technique of Wigner Distribution. The main idea is that there exists a certain level of any noise in sample considered in real experiment. This noise has been related to the fluctuation of density of matter or number of particles (i.e., $\frac{\delta\rho}{\rho}$ or $\frac{\delta N}{N}$) which ultimately might be related to the inevitable quantum noise. Now a real experiment in the laboratory can be done using different levels of noise externally and find the continuous reduction of the coherence of interfering beams.

In fact, Rauch et al.[Rauch 1996] are preparing for this sort of experiment in near future. This result has great significance not only from the point of view of real laboratory conditions but also form the point of view of restoring the age old debate concerning the measurement process itself.

Let us briefly discuss the main mathematical steps behind this approach [Rauch, 1995]. The wave function of the beam behind the interferometer in forward direction can be constructed from the beams of the path I and II. They are equal in amplitude and phase. Now if we insert a sample in one of the coherent beams, it causes a phase shift χ which is determined by its index of refraction n and its thickness D , then

$$I_0 \propto \mid \Psi_0^I + \Psi_0^{II} \mid^2 \propto 1 + \cos\chi \tag{2.45}$$

with

$$\chi = (n-1)kD = -Nb_c\lambda D = \Delta.\vec{k} \tag{2.46}$$

Here, N denotes the particle density, b_c the coherent scattering length of the sample and Δ can be interpreted as the spatial shift of the wave in relation to the reference beam. The direction of this shift depends on the orientation of the phase shifter to the beam.

The wave function can also be used to define the Wigner momentum-position quasi-probability function, given by

$$W(\vec{r},\vec{k}) \propto \int \Psi^*(\vec{r}+\frac{\vec{r'}}{2})\Psi(\vec{r}-\frac{\vec{r'}}{2})e^{i\vec{k}.\vec{r}}d^3r' \tag{2.47}$$

In case of neutron interferometry, the wave function to be used for above Wigner function is as

$$\Psi_s(\vec{r},\Delta) + \Psi(\vec{r}+\Delta)$$

After some calculations we have

$$\begin{aligned} W_s(x,k,\Delta) &= W(x,k) + W(x+\Delta,k) + 2W(x+\tfrac{\Delta}{2},k)\cos(\Delta k) \\ &\propto \exp[-\tfrac{(k-k_0)^2}{2\delta k^2}]\{\exp(-\tfrac{x^2}{2\delta x^2}) + \exp[-\tfrac{(x+\Delta)^2}{2\delta x^2}] \\ &+ 2\exp[-\tfrac{(x+\frac{\Delta}{2})^2}{2\delta x^2}]\cos(\Delta.k)\} \end{aligned} \tag{2.48}$$

Thus, the Wigner function of the coherent Schrödinger cat-like states becomes the sum of the Wigner function of the two shifted wave packets and a cross term oscillating strongly in the case of increasing phase shifts. Now any physical system has intrinsic fluctuations and inhomogeneities around certain mean values. The roughness of the surfaces causes variations of the thickness. Density fluctuations inside a phase shifter arise due to thermodynamical reasons and residual stresses. Thus, imperfections exist even at zero temperature, and even a magnetic field used to as phase shifter exhibits intrinsic photon numbers and therefore, fluctuations. Let us now discuss the effects of this fluctuations and inhomogeneity (with dimensions larger than the coherence lengths of the neutron beam) in the interference phenomena.

The phase shifter may have variations of its thickness δD around its mean value D_0 and may have density variations δN around N_0 which influences the phase shift. In case of Gaussian distribution functions and $\delta D \ll D_0$ and $\delta N \ll N_0$, the above Wigner function takes the form

$$\begin{aligned} \overline{W}_s(x,k,\Delta) &= \int W(D)W(N)W_s[x,k,\Delta(D,N)]dDdN \\ &\propto \exp[-\tfrac{(k-k_0)^2}{2\delta k^2}]\{\exp[-2(\delta k\ x^2)^2] \\ &+ (1+\epsilon)^{-\frac{1}{2}}\exp[-\tfrac{2\delta k^2(x+\bar{\Delta})^2}{(1+\epsilon)}] \\ &+ 2(1+\tfrac{\epsilon}{4})^{-\frac{1}{2}}\exp[-2\delta k^2\tfrac{(x+\frac{\bar{\Delta}}{2\epsilon})^2}{(1+\frac{\epsilon}{4})}] \\ &\times \exp[-(\tfrac{\epsilon}{4})(1+\tfrac{\epsilon}{4})(\tfrac{k^2}{2\delta k^2})]\cos[k\tfrac{(\frac{x\epsilon}{2}-\bar{\Delta})}{(1+\frac{\epsilon}{4})}] \end{aligned} \tag{2.49}$$

where

$$\bar{\Delta} = \Delta_0(\frac{k_0}{k})$$

and

$$\epsilon = 4(\bar{\Delta}\delta k)^2[(\frac{\delta D}{D_0})^2 + (\frac{\delta N}{N_0})^2]$$

which gives rise the additional damping factor for the interference pattern

$$\begin{aligned} I(\Delta_0) &= \int \bar{W}_s(x,k,\Delta)dkdx \\ &\propto 1+\exp[-\tfrac{(\Delta_0\delta k)^2}{2}] \\ &\times \exp\{-[(\tfrac{\delta D}{D_0})^2+(\tfrac{\delta N}{N_0})^2]\tfrac{(\Delta_0 k_0)^2}{2}\} \\ &\times \cos(k_0\Delta_0) \end{aligned} \tag{2.50}$$

It is important to note that the cross term of the Wigner function is more influenced by these damping factors rather than Wigner function of the individual seperated packets. It can be shown that fluctuations wash out the cross term in any case, which demonstrates that the coherent seperation of a Schrödinger cat-like state becomes progressively more difficult with increasing seperation because even small fluctuations have to be avoided. It would be very interesting to find the effect of unavoidable zero-point fluctuations.

This approach can be directly related to the proposal of decoherence within the framework of fluctuating metric [Sanchez-Gomez 1995]. Before discussing the role of space-time metric fluctuation in decoherence problem, let us begin with the stochastic mechanism in general for reduction of wave packet.

Ghirardi, Pearle and Rimini [Ghirardi 1990] studied a Continuous Spontaneous Localization (CSL) approach in Hilbert space framework. A stochastic evolution equation of the Ito form is assumed

$$d|\phi> = [-(1/2)^r (A-R)^2 dt + (A-R)dB]|\phi> \tag{2.51}$$

where, A is a set of commuting self adjoint operators, $R = <\phi|A|\phi>$ and dB satisfies Wiener process

$$\overline{dB_i} = 0 \tag{2.52}$$

$$\overline{dB_i}\,\overline{dB_j} = \delta_{ij} r dt \tag{2.53}$$

Essentially, quantum mechanics is modified by superimposing a Markov process in Hilbert space upon the Schrödinger evolution. This gives rise to a continuous localization of the wave equation. The decoherence is obtained through density matrix of "classical" systems. Sanchez-Gomez et.al. [Sanchez-Gomez 1995] rightly pointed out :

How one can get such a typical process that could give rise to the aforementioned Markov process ?

In fact, if we closely look into CSL model then one can find two stochastic processes : one linear and the other non-linear one. He associated this linear one with metric fluctuation and nonlinear one with the particle's gravitational field.

Diosi and collaborators [Diosi 1993] studied the damping of coherence considering the existence of universal fluctuations of the gravitational field (of the vacuum). In the Newtonian limit of general relativity, the metric tensor g_{ij} has the following non-zero components :

$$g_{ij} = c^2 + 2\phi \quad \text{for } i = 0 \tag{2.54}$$

$$= -1 \quad \text{for } i = 1,2,3$$

where, ϕ is Newtonian gravitational potential and c the velocity of light and $|\phi| << c^2$ is assumed. ϕ is considered as stochastic variable with the following statistical properties :

$$< \phi(\vec{r}, t) >= 0$$

$$< \phi(\vec{r}, t)\phi(\vec{r}', t') >= G|\vec{r} - \vec{r}'|^{-1}\delta(t - t') \tag{2.55}$$

ϕ being treated as Gaussian white noise.

Following this, a master equation can be derived with such a gravitational noise. This noise produces a damping term in the master equation. Total Hamiltonian can be written as

$$H = \hat{H}_0 + \int \phi(\vec{r}, t)\hat{f}(\vec{r})d^3\vec{r} \tag{2.56}$$

where, $\hat{H}_0$ is the non-gravitational part of the Hamiltonian, $\widehat{f}(r)$ stands for the operator of the local mass density of the system. In this stage, we can construct a master equation for density matrix ρ in coordinate representation using Feynman-Vernon's method. It can be written as

$$\frac{\partial}{\partial t}\rho(\vec{x}, \vec{x}', t) = \mathcal{L}[\rho] - g(\vec{x}, \vec{x}')\rho(\vec{x}, \vec{x}', t) \tag{2.57}$$

the operator, here, $\mathcal{L}$ represents the quantum evolution and is given by (also in coordinate representation)

$$\mathcal{L}\ [\rho] = -\frac{i\hbar}{2m}[\Delta_x - \Delta_{x'}] < \vec{x}|\rho|\vec{x}' > -\frac{i}{\hbar}[U(\vec{x}, t) - U(\vec{x}', t)] < \vec{x}|\rho|\vec{x}' > \tag{2.58}$$

U is any external deterministic potential. For a free particle $U = 0$. The decoherence part can be written using stochastic gravitational potential $\phi(\vec{x}', t)$, as

$$g(\vec{x}, \vec{x}') = \frac{\tau}{\hbar^2}\left[< V^2 > - < V(\vec{x}, t)V(\vec{x}', t) >\right] \tag{2.59}$$

with $V(\vec{x}, t) = \frac{1}{2}mc^2\phi(\vec{x}, t)$, τ denoting the correlation time. It is assumed that τ is much smaller than the relevant quantum time of the system. For example, a typical colloid grain is of the order of 10^{-5} cm. Let us take $L = 10^{-4}$ cm, then $\tau = 3 \times 10^{-15}$ s. The corresponding quantum time is $= \frac{mL^2}{\hbar}$, which becomes (for proton) $\approx 10^{-9}$ s. Hence our above assumption holds. We can calculate the function g considering the correlation function for $\phi(\vec{x},t)$. By approximating the space part as

$$\sigma^2[1-\frac{(\vec{x}-\vec{x}')^2}{L^2}] \quad \text{for} \quad |\vec{x}-\vec{x}'|<L \tag{2.60}$$

and

$$0 \qquad \text{for} \qquad |\vec{x}-\vec{x}'|>L$$

the master equation follows as

$$\frac{\partial}{\partial t}\rho(\vec{x},\vec{x}',t)=\mathcal{L}[\rho]-\left(\frac{Gm^2}{4\hbar L^3}\right)(\vec{x}-\vec{x}')^2\rho(\vec{x},\vec{x}',t) \text{ for } |\vec{x}-\vec{x}'<L \tag{2.61}$$

$$=\mathcal{L}[\rho]-\left(\frac{Gm^2}{4\hbar L}\right)\rho(\vec{x},\vec{x}',t) \text{ for } |\vec{x}-\vec{x}'|>L \tag{2.62}$$

It is now evident that the non-unitary term is negligible in case of microscopic case. For example, $L=10^{-4}$cm and by taking such a system to be a proton, we get

$$\lambda_m=\frac{Gm^2}{4\hbar L^3}\simeq 10^{-17}\text{cm}^{-2}s^{-1} \tag{2.63}$$

Hence, the relative coherence between parts of the proton wave packet that are separated by up to 1 cm would survive up to $\sim 10^{24}$s. This indicates that the time evolution is practically unitary. Now, this framework for microscopic case can be generalized to macroscopic case, the stochastic potential will be given by

$$V(\vec{x}_1,\cdots\vec{x}_N,t)=\frac{1}{2}mc^2\sum_i\phi_i(\vec{x}_i,t) \tag{2.64}$$

For macroscopic case, $R>>L$, in which, R is radius of the body, the center of mass can be factored out. Here L is the correlation length. After simple calculations, we can write the master equation as

$$\frac{\partial}{\partial t}\rho_0(\vec{x}_0,\vec{x}_0',t)=\mathcal{L}[\rho_0]-\lambda_0(\vec{x}_0-\vec{x}_0')^2\rho_0(\vec{x}_0,\vec{x}_0',t) \text{ for } |\vec{x}_0-\vec{x}_0'<L \tag{2.65}$$

$$=\mathcal{L}[\rho_0]-\gamma_0\rho_0(\vec{x}_0,\vec{x}_0',t) \text{ for } |\vec{x}_0-\vec{x}_0'|>L \tag{2.66}$$

Here, ρ_0 represents the density matrix for center of mass quantum state and

$$\lambda_0=\frac{GMD}{\hbar}; \qquad \gamma_0=\lambda_0 L^2 \tag{2.67}$$

M and D being the mass and the density of the body respectively.

The above master equation clearly indicates the decoherence between the off-diagonal terms of the center of mass density matrix of a macroscopic body. Let us consider, say, $M = 1\text{gm}$; $D = 1\text{g/cm}^3$. Then the correlations between macroscopic quantum states separated by more than $L(10^{-4}$ cm, say) would decay in about 10^{-12}s. It should be noted that this decoherence is produced by stochastic fluctuation of gravitational field (in vacuum). This is supposed to be universal fluctuation and acts like an environment for any body.

It is worth mentioning that Karolyhazy [Karolyhazy 1991] (K-model) proposed a framework in which the stochastic fluctuation of space-time metric can give rise to break down of quantum superposition principle in case of macroscopic systems.

Hazy Space-Time of Karolyhazy Model :

The principal idea behind Karolyhazy [Karolyhazy 1991] model lies on the fact that the quantum mechanical uncertainties in the positions and momentum of material bodies induce a minimal unavoidable uncertainty in the length of time intervals and hence the loss of classical space-time to the same extent. Quantum states will then propagate on a slightly hazy space-time. Karolyhazy emphasized that the amount of uncertainty incorporated into the structure of classical space-time is the apporopriate amount needed to destroy the coherence of the quantum states of macroscopic bodies, whereas the coherence of the states of microsystems will be practically unaffected [Frenkel 1995]. In this approach the minimal quantum mechanical uncertainty $(\Delta_Q T)_{\text{min}}$ of a time interval of length T is related to T by the formula

$$(\Delta_Q T)_{\text{min}} \approx (\Lambda/c)^{\frac{2}{3}} T^{\frac{1}{3}} \tag{2.68}$$

where $\Lambda = \sqrt{\frac{G\hbar}{c^3}} \approx 10^{-33} cm$ is the Planck length. Here, $\Delta_Q T$ denotes not only the quantum uncertainty Δx in the position of a body, but also the uncertainty $\Delta K = v\Delta p$ in the kinetic energy. This ΔK contribute to the uncertainty of the structure of space-time and consequently of the length of T. G (i.e., gravitational constant) appears in the expression of this contribution. Again, as $\Delta X.\Delta p \geq \frac{\hbar}{2}$, ΔX and ΔK work against each other. That is why a minimum of $\Delta_Q T$ exists. This $(\Delta_Q T)_{\text{min}}$ can be derived by several ways [Frankel. 1995]. We shall concentrate our discussions on three main ways of derivation as follows:

- (a) The above expression for $(\Delta_Q T)_{\text{min}}$ can be deduced for time intervals along world lines of slowly moving (or standing) bodies. The reference frame in which the bodies should move slowly is the one in which the $2.7^0 K$ microwave background radiation is isotropic. Since the K-model is a non-relativistic quantum mechanical model, this is an adequate limitation.

- **(b)** The expression of $(\Delta_Q T)_{\min}$ is valid for the time intervals T such that

$$T \gg T_{\text{Planck}} = \frac{\Lambda}{c} \approx 10^{-43} \text{ sec}$$

This is obvious, as $T \leq T_{\text{Planck}}$, the very concept of classical space-time becomes questionable and in that case one should not attribute physical meaning to the above expression of $(\Delta_Q T)_{\min}$.

- **(c)** In order to reach the minimal uncertainty $(\Delta_Q T)_{\min}$, one should allow for bodies of unrealistically high density. For realistic bodies $\Delta_Q T$ is many orders of magnitude larger than $(\Delta_Q T)_{\min}$. However, in the K-model only the existence of the theoretical lower bound $(\Delta_Q T)_{\min}$ is exploited, there is no need to reach it in experiments. and for $T = 1$ sec, ;

$$(\Delta_Q T)_{\min} \sim 10^{-29} \text{sec}$$

Let us now consider how the indeterministic element is introduced in the time evolution of quantum states. It should be noted that the very existence of the quantum mechanical spreads $\Delta x, \Delta p$, $\Delta_Q T$ of various quantities do not lead to a breakdown of the deterministic Schrödinger evolution of the quantum state of a closed system. In the GRW model, the indeterministic element is introduced in the form of sponteneous random localization. However, in the K-model, the indeterminacy is introduced primarily not into the time evolution of the state vector but into the structure of classical space-time. But if we look into the expression of $(\Delta_Q T)_{\min}$, it clearly indicates that it does not contain any parameter of particular material body other than the constants of nature. This suggests a general property of time intervals. Loosely speaking, quantum mechanics and classical general relativity, taken together, produce in the length T of any time interval of a spread at least as large as $(\Delta_Q T)_{\min}$. Karolyhazy studied the influence of the haziness of space-time on the quantum states of various isolated systems in the following way:

Let a world event, which in Minkowskian space-time would occur at the point $(\vec{X}, t)$, at an uncertain moment

$$t_\beta(\vec{X}, t) := t + \tau_\beta(\vec{X}, t)$$

Here, τ_β constitute a random set of $(\vec{X}, t)$ function; β labels the different functions of the set.

Let $\vec{x}, t : \vec{x}' = \vec{x}, t' > t$ is a zero velocity world line segment. The time interval $T = t' - t$ goes over into

$$T_\beta = t_{\beta'} - t_\beta = T + \tau_\beta(\vec{X}, t') - \tau_\beta(\vec{X}, t) \tag{2.69}$$

Taking $\langle \tau_\beta(\vec{X}, t) \rangle = 0$, we get

$$\langle T_\beta \rangle = T \tag{2.70}$$

where $\langle \cdots \rangle$ stands for an average over the set labelled by β. This implies that on average, the time intervals have their Minkowskian value. Then,

$$\Delta T_\beta := T_\beta - \langle T_\beta \rangle \tag{2.71}$$

$$\begin{aligned} \langle (\Delta T_\beta)^2 \rangle &= \langle (T_\beta - \langle T_\beta \rangle)^2 \rangle \\ &= \langle (\tau_\beta(\vec{X}, t') - \tau_\beta(\vec{X}, t))^2 \rangle \end{aligned} \tag{2.72}$$

Identifying $\sqrt{(\Delta \tau_\beta)^2}$ with ΔT , we obtain

$$\langle (\tau_\beta(\vec{X}, t') - \tau_\beta(\vec{X}, t))^2 \rangle \approx (\frac{1}{c})^{\frac{4}{3}} \mid t' - t \mid^{\frac{2}{3}} \tag{2.73}$$

It is worthwhile to note that $\langle (\tau_\beta(\vec{X}, t') - \tau_\beta(\vec{X}, t))^2 \rangle$ is given along $\vec{X}' = \vec{X} =$ constant worldlines. But if we consider this function on space-like surfaces $t = t' =$ constant, a kind of indeterminacy will be introduced on the evolution of quantum states. Taking the isotropic and homogeneous nature of space-time, the above function has been shown to depend only on the differences :

$$r := \mid \vec{X}' - \vec{X} \mid; \qquad \theta := \mid t' - t \mid \tag{2.74}$$

Now, on space-like surfaces i.e., $t' = t =$constant, the above function can depend only on r. So,

$$\langle (\tau_\beta(\vec{X}, t') - \tau_\beta(\vec{X}, t))^2 \rangle \approx \frac{\Lambda^{\frac{4}{3}}}{c^2} \mid \vec{X}' - \vec{X} \mid^{\frac{2}{3}} \tag{2.75}$$

Let us consider an isolated system of N quantum particles with masses $M_1, M_2, \cdots, M_N$. In nonrelativistic quantum mechanics, the rest energy phase factor

$$\exp[-\frac{i}{\hbar}\sum_{i=1}^{N} M_i c^2 t)] := \exp(i\Phi(t)) \tag{2.76}$$

of the wave function $\Psi(\vec{X}, t)$ of the system is usually omitted. As this phase factor is independent of $x := [\vec{X}_1, \cdots, \vec{X}_N]$ and $p := [-i\nabla_1, \cdots, -i\nabla_N]$, this phase factor would drop out from all observables. However, with

$$t \to t + \tau_\beta(\vec{X}, t); \quad M_i c^2 \to M_i c^2 \left[t + \tau_\beta(\vec{X}_i, t)\right]$$

As a result

$$\exp(i\Phi(t)) \to \exp(i\Phi(t)). \exp(i\Phi_\beta(x, t)) \tag{2.77}$$

where

$$\Phi_\beta(x, t) := -\frac{c^2}{\hbar}\sum_{i=1}^{N} \tau_\beta(\vec{x}_i, t)$$

The first factor i.e., containing $\Phi(t)$ can be omitted but the second factor i.e., $\exp(i\Phi_\beta(x, t))$ should be kept as it depends on x. Hence, a pure quantum state of an isolated system of N quantum particles is not a usual Schrödinger wave equation $\Psi(x, t)$, but $\Phi(x, t)$ multiplied by uncertain x-dependent phase factor $\exp(i\Phi_\beta(x, t))$. Here, the state is not one of the functions

$$\Psi_\beta(x, t) := \exp(i\Phi_\beta(x, t))\Psi(x, t) \tag{2.78}$$

with same probability p_β. This means that the state is a usual mixture. The uncertainty would then not in the state, but in our knowledge of the state. In the phase of a pure K-model state, the uncertainty is "objective". The pure K-state of an isolated system is the collection

$$\{\Psi_\beta(x, t)\} = \{\exp(i\Phi_\beta(x, t))\}\Psi(x, t) \tag{2.79}$$

and not the mixture of the functions Ψ_β. Only the collection (and not its individual members) contains the full information about the state. In particular, it carries the information about the amount of uncertainty of the relative phase of $\{\Psi\}$ between pairs of points of the configuration space. Tha main difference between a pure Schrödinger state Ψ and the pure $\{\Psi_\beta\}$ state corresponding to it is that while a Schrödinger state is always perfectly coherent (in the sense that its relative phase between any pairs of points x, x' is uniquely determined) a $\{\Psi_\beta\}$ state is never perfectly coherent, as the amount of uncertainty of its relative phase is never zero.

However, as the uncertain phase factor is symmetric under the permutations of N-particles, so that a K-state has the same symmetry as the corresponding Schrödinger state. Again, the substitution $t \to t + \tau_\beta$ should have been carried out not only in the rest energy phase factor, but in the Schrödinger wave function

$$\Psi(x,t) = \exp[(-\frac{i}{\hbar}\hat{H}t)]\Psi(x,0) \tag{2.80}$$

The matrix elements of the Hamiltonian $\hat{H}$ in non-relativistic cases are much smaller than the rest energies of the particles.

Now the relative phase of a member Ψ_β of a K-state at two points x and x' is

$$\begin{aligned} \delta_\beta(x,x',t): &= \Phi_\beta(x',t) + \arg\Psi(x',t) - \Phi_\beta(x,t) - \arg\Psi(x,t) && (2.81)\\ \langle\delta_\beta(x,x',t)\rangle &= \arg\Psi(x',t) - \arg\Psi(x,t) && (2.82)\\ \Delta\delta_\beta(x,x',t) &= \Phi_\beta(x',t) - \Phi_\beta(x,t) && (2.83)\end{aligned}$$

After some calculations, we get a time independent expression for the spread of the relative phase

$$\Delta^2_\phi(x,x') \approx \Lambda^{\frac{4}{3}}\frac{c^2}{\hbar^2}\sum_{i,l=1}^{N} M_i M_i [2 \mid X_i' - X_l \mid^{\frac{2}{3}} - \mid X_i' - X_l' \mid^{\frac{2}{3}} - \mid X_i - X_l \mid^{\frac{2}{3}}] \tag{2.84}$$

where $\Delta^2_\phi(x,x') := \langle(\Delta\delta_\beta(x,x',t)^2\rangle$.

The above expression can also be written as

$$\Delta^2_\phi(x,x') \approx \Lambda^{\frac{4}{3}}\frac{c^2}{\hbar^2}\int\frac{dk}{k^{\frac{1}{3}}}[\mu_k(x') - \mu_k(x)]^2 \tag{2.85}$$

where $\mu_k(x) := \sum_{l=1}^{N} M_l\ e^{ikx_i}$ is the Fourier transform of the mass distributions of the N-particles with masses $M_1, \cdots, M_N$ in the configuration $x = [\vec{X}_1, \cdots, \vec{X}_N]$,

It is now clear that for a physical system $\Delta\Phi$ increases with the seperation $\mid \vec{X}_i' - \vec{X}_l \mid$. As a result, the coherence of $\{\Psi_\beta\}$ detoriates if Ψ expands in the configuration space.

Let us now characterize the degree of coherence of a given state $\{\Psi_\beta\}$ of a given physical system. This can be done mainly in the two ways.

– (a) If in $\{\Psi_\beta\}$ the wave function occupies only a domain Ω such that

$$\Delta_\phi(x,x') \ll \pi \quad \text{for all} \quad x,x' \in \Omega$$

Then, in good approximation, in Ω

$$\Phi_\beta(x',t) = \Phi_\beta(x,t) \quad \text{for all } \beta \tag{2.86}$$

Φ'_βs are practically independent of x in Ω. This implies that $\{\Psi_\beta\}$ is practically equivalent to the ray $\{\exp(i\Phi_\beta(t)).\Psi(x,t)\}$ i.e., K-state is practically equivalent to the usual pure state Ψ. Here, the coherence of the K-state is nearly perfect.

– (b) If Ψ occupies a domain containing subdomains Ω, Ω' of considerable weights such that

$$\Delta_\Phi(x,x') > \pi$$

for all $x \in \Omega$, and for all $x' \in \Omega'$ then the relative phases between these subdomains are completely uncertain.

The coherence between the components of $\{\Psi_\beta\}$ belonging to these subdomains is lost. $\{\Psi_\beta\}$ is far from being equivalent to a single ray. However, within subdomains inside which $\Delta_\Phi < \pi$ a certain degree of coherence persists, and inside suffuciently small subdomains the coherence is nearly perfect. a maximal domain of the configuration space inside which $\Delta_\phi < \pi$ may be called a "coherence cell". Its extent varies from system to system.

For a single microparticle, say an electron, the linear size of the coherence cell is of the order of $10^{33}cm$ in a domain occupied by any realistic wave function of a microparticle $\Delta_\phi \ll \pi$. Therefore the K-state of an isolated microparticle is practically equivalent to a usual pure quantum state.

In case of a macroscopic body, the value of Δ_ϕ can be determined by the extent of the wave function Ψ in the centre of mass coordinate subspace of the configuration space. Let us consider a body of 1gm, whose corresponding linear size of the coherence cell is of the order of10^{-16}cm. So a K-state of the isolated body in which Ψ extends over a domain with linear size $2a_c$ in the centre of mass subspace, contains incoherent components of considerable weights. Frenkel [Frankel 1995] discussed three different realistic cases in K-models regarding the time evolution of state vectors.

(i) In case of an isolated microparticle, a coherence cell is so large that $\Delta_\Phi \ll \pi$. Therefore no random localizatiion of Ψ occurs and the Schrödinger equation holds without restriction.

(ii) For isolated macroscopic body, the wave function can expand to a domain larger than a coherence cell. If at some moment of time $\Psi(x,t)$ occupies in the center of mass coordinate subspace a single coherence cell of linear dimension a_c, then after a time

$$\tau_c \approx \frac{Ma_c^2}{\hbar}$$

it will occupy a domain of linear size $2a_c$.

Let us consider a body of mass $1gm, a_c \approx 10^{-16}cm$, so that $\tau_c \approx 10^{-5}sec$. The localization-expansion cycles, at a rate $\tau_c^{-1} = 10^5$ localization/sec, keep the quantum mechanical spread of the position of the center of mass of a macroscopic body practically point like. In this respect the body behaves classically.

The mathematical description of the random localization in K-model is similar to that of GRW model for finite cycle periods. A detail comparison of the GRW model with K-model is done by Frenkel [Frenkel 1990].

(iii) K-model has been applied to study gases as a cloud chamber filled with vapour.

Continuous localization has also been studied by Karolyhazy. It seems from the above analysis that Karolyhazy model starts with some intuitive notion and able to describe decoherence in a more realistic way than the other localization models. But Diosi and Lukacs [Diosi 1987, 1993] pointed out some unpleasant feature of this K-model. This model predicts a very large mean value of the space-time curvature and such a curvature would imply a mean energy density of the Universe much larger than that of a neutron star. However, if we consider the conformal metric fluctuation, this unpleasant feature does not arise.

The non-unitary term in the framework of conformal metric fluctuation is responsible for the change of energy with time, however, isolated be the system. This is of course negligible but it is always positive. Obviously it is an unpleasant feature of this model. Moreover,

$$\frac{d}{dt} < E_{cm} >= 2GD\hbar \simeq 10^{-41}W \tag{2.87}$$

which is independent of the mass.

This constant increase of energy may be due to the fact that the vacuum fluctuations act to the particle in a non-dissipative way but the particle's reaction on the metric is disregarded. But this reaction may be relevant for "mesoscopic" system i.e., $10^{-16}\text{gm} < M < 10^{-8}\text{gm}$. The mesoscopic case is worth considering since this kind of behavior in between the quantum and classical ones, which perhaps could give rise to some experimental tests of the present model. This model predicts the said mesoscopic region to lie within the range $10^{-16}\text{gm} < M < 10^{-8}\text{gm}$. The future work using mesoscopic systems may give rise to the new test of this model. However, we shall discuss other aspects of conformal metric fluctuations and its implications in cosmological scenario in chapter 4.

2.4. Frame Dependence of Scattering Matrix :

Ingraham [Ingraham 1985] considered the role of Lorentz transformations in stochastic space-time. He discussed in details the Lorentz transformations among different stochastic observers and the way to calculate the mean values. This kind of stochasticity may be manifested in very high energy scattering cross-sections. To calculate the high energy cross-sections it is necessary to consider the relativistic framework for stochastic space-time. We have already mentioned that four dimensional probability for stochastic metric $g_{\mu\nu}$ can be made invariant by introducing a time-like vector n. This time like vector can be attached to a fixed frame of reference like aether in older time. In that case, the scattering matrix or S- matrix operator would depend on the parameters of one fixed frame and would immediately violate the equivalence of all Lorentz frames. Therefore, the time-like vectors n must be allowed to vary over the whole set of Lorentz frames. It immediately raises the question regarding the physical meaning of frame dependence. Ingraham put forward a hypothesis regarding this frame dependence. The frame dependence means the frame in which the experiment is actually done, the measurements (involving detectors, collimators etc. fixed in that frame) actually performed. Let

$$|K> \equiv |k_1, k_2, \cdots, k_n> \quad \text{and} \quad |Q> \equiv |q_1, q_2, \cdots, q_m>$$

be $n-$ and $m-$ free meson states with four momenta $k_1, k_2, \cdots, k_n$ and $q_1, q_2, \cdots, q_m$ referred to the unprimed frame $\mathcal{L}$. Also let us take

$$|K'> \equiv |k'_1, k'_2, \cdots, k'_n> \quad \text{and} \quad |\, Q'> \equiv |\, q'_1, q'_2, \cdots, q'_m>$$

as $n-$ and $m-$ free meson states with four-momenta $k'_1, k'_2, \cdots, k'_n$ and $q'_1, q'_2, \cdots, q'_m$ referred to the primed frame $\mathcal{L}'$. Then, if $|\, K>' \equiv |\, K'>'|_{k'=k}$ and similarly for $|\, Q>'$, the matrix elements

$$< Q \mid S \mid K > \text{ and } < Q \mid S \mid K >' \tag{2.88}$$

are possibly the amplitudes for the same experiment n mesons $\rightarrow$ m mesons as performed by $\mathcal{L}$ and $\mathcal{L}'$. Now if the scattering matrix depends on frames, the matrix elements become

$$< Q \mid S(\mathcal{L}) \mid K > \text{ and } < Q \mid S(\mathcal{L})') \mid K >'$$

It should be mentioned that in QED or in particle physics, there is no such frame dependence concept for S matrix. We shall show that it might be related to the time-like vector as considered in relativistic stochastic metric tensor. As for parameters of a frame $\mathcal{L}$ we are thinking of the unit time-like vectors $\vec{n}(\mathcal{L})$ directed along $\mathcal{L}$'s positive time axis i.e.,

$$\begin{aligned} \vec{n}^2(\mathcal{L}) &= -1, \ n^o(\mathcal{L}) > 0 \\ n^\mu(\mathcal{L}) &= (1,0,0,0) \end{aligned} \tag{2.89}$$

Then the generalized transformation law for the S-operator takes the form

$$S(L^{-1}\mathcal{L}) = U(L)S(\mathcal{L})U(L)^{-1} \tag{2.90}$$

where, $x' = LX$ and L denotes Poincaré transformation which bodily rotates or translates the whole physical system in space-time.

Again we know that S-operator is a functional of the field and it must be the same functional form for all Lorentz frames, otherwise their equivalence is violated. But the question arises whether this frame dependence is compatible with established physical principles. Ingraham discussed these in details in one of his recent papers [Ingraham 1985]. The interesting aspect of this frame dependence is that the meson propagator $< 0 \mid T\phi(x,\mathcal{L})\phi(x\prime,\mathcal{L}') \mid 0 >$, in momentum space becomes

$$D^c(q,\mathcal{L}) = D^c(q)|g|^2 = \frac{|g(q.n(\mathcal{L}))|^2}{q^2+\mu^2-i\epsilon} \tag{2.91}$$

The form factor g can introduce an ultraviolet cut-off into the S-operator. In its simplest form, g could be

$$g(q.n(\mathcal{L})) = \theta(1-q_\perp^2(\mathcal{L})l^2) \tag{2.92}$$

with $q_\perp{}^2(\mathcal{L}) = q^2 + [q.n(\mathcal{L})]^2$ and $q_\perp{}^2(\mathcal{L})$ as the square of the spatial part of the four q in frame $\mathcal{L}$. Other type of cut-off may also be used.

We like to emphasize, here, that if we consider stochastic space-time i.e., if space-time coordinates are be treated as random, such a frame dependence is inevitable.

2.5. E-P-R- Correlations :

In our framework, the variables ν play the role of hidden degree of freedom. These hidden degrees of freedom can also be related to the hidden variables in de Broglie-Bohm-Vigier hidden variable interpretation of Quantum mechanics. These variables help us to understand the E-P-R like correlation as follows :
Let us consider the total wave function for the two particles 1 and 2 as

$$\Psi(x_1, x_2, \nu) = \int \exp[\frac{2\pi i}{h}(x_1 - x_2 + x_0)p + \nu(x_1) + \nu(x_2)]dp \tag{2.93}$$

and $\Psi(x_1, x_2) = \int \Psi(x_1, x_2, \nu)d\nu$.
This is the usual wave function of the E-P-R paradox. Then,

$$\frac{\hbar}{i}\frac{\partial}{\partial x_1}\Psi_p(x, \nu) = [p + \frac{\partial \nu}{\partial x_1}]\Psi(x, \nu) \tag{2.94}$$

and

$$\frac{\hbar}{i}\frac{\partial}{\partial x_2}\Psi_p(x, \nu) = \left[-p + \frac{\partial \nu}{\partial x_2}\right]\Psi_p(x, \nu) \tag{2.95}$$

for particle 1 and 2 respectively. So,

$$p_2 = -p_1 + \left[\frac{\partial \nu}{\partial x_1} + \frac{\partial \nu}{\partial x_2}\right] \tag{2.96}$$

where p_1 and p_2 are the eigenvalues of P_1 and P_2 respectively so that the momentum of the 2nd particle can be inferred by measuring that of the 1st particle only if

$$\frac{\partial \nu}{\partial x_1} + \frac{\partial \nu}{\partial x_2} = 0 \text{ or } \frac{\partial \nu}{\partial x_1} = -\frac{\partial \nu}{\partial x_2} \tag{2.97}$$

Then the paradoxical situation can be resolved within the context of a generalized interpretation of quantum mechanics, if we consider the relativistic Brownian motion in Finsler space taking the directional variable as hidden parameters.

The role of these hidden variables is to correlate the object with the instruments as well as with other objects. Then, as soon as the two particles are correlated to form (say) a singlet state, then these hidden parameters correlate the two particles in such a way that they possess a definite distribution of directional variables. Now, once their distribution is fixed with respect to the other; it can not be changed since they are responsible for the internal structure of the particles so that when we try to measure the variables like spin etc. we have to arrange the apparatus in such a way that a correlation happens to exist between the object and the instrument. As a result the apparatus for the 2nd particle should be arranged in such a way so as to have the desired correlation. Of course, here also, the arrangement of the 1st measurement does not disturb anyway that of the 2nd apparatus. In this way the incompleteness argument put forward by the E-P-R statement can be completed by considering the complete wave function.

2.6. 4π- Periodicity of Spinor Wave Function :

The generalization of quantum mechanics has also been proposed recently [Ktorides 1980] considering the violation of $SU(2)$ symmetry within the range of strong interaction.Ktorides et al. [Ktorides 1980] studied whether any deviation from the conventional value of the spin of hadrons under strong interactions is conceivable and quantitatively treatable. This possible deviation from the value of the spin implies a corresponding deviation from Pauli's principle. The starting point is to consider the massive, spinning and charged particle(as characterized under electromagnetic interactions and Poincaré symmetry) in the presence of strong interactions. The spin as well as other intrinsic characteristics of extended, massive particles are supposed to be mutated under strong interactions at distances smaller than their charge radius. These dynamical effects can apparently be conjectured to account for the non point-like nature of particles, their necessary state of penetration to activate the strong interactions and the consequential emergence of broader forces which imply the breaking of $SU(2)$ -spin symmetry. Here, the characterization of the mutated value of the spin is made via the transition from the associative enveloping algebra of $SU(2)$ to a nonassociative Lie-admissible form. The departure from the original associative product then becomes directly representative of the breaking of the $SU(2)$-spin symmetry, the presence of forces more general than those derivable from a potential and the mutated value of the spin. Such a departure of the spin from conventional quantum mechanical values implies the inapplicability of Pauli's exclusion principle under strong interactions, because, according to this hypothesis, particles that are fermions under long-range electromagnetic interactions are no longer fermions under these broader, short range, forces.

The salient aspect of this studies can be briefly summerized as follows :

The Galilei and Einstein notions of particles essentially characterize a particle under strong interactions, that is, with forces derivable from a potential (as typical of the electromagnetic interactions), via an articulated body of compatible mathematical tools of analytic, algebraic, and geometric character.

At classical discrete level, these tools are expressible via Hamilton equations without external terms with underlying Lie algebra and symplectic structurte, i.e.,

$$\dot{a}^{\mu} = \omega^{\mu\nu}\frac{\partial H}{\partial a^{\nu}} = 0; \quad a = (r, p) \tag{2.98}$$

$$\begin{aligned} H &= T + V, \quad \mu = 1, 2, \cdots, 2n \\ [A, B]_{\mathrm{cl}} &= \frac{\partial A}{\partial r_k}\frac{\partial B}{\partial p_k} - \frac{\partial B}{\partial r_k}\frac{\partial A}{\partial p_k} \\ \omega_2 &= dR_1^0 \\ \omega_{\mu\nu} &= \frac{\partial R_{\nu}^0}{\partial a^{\mu}} - \frac{\partial R_{\mu}^0}{\partial a^{\nu}}; \quad R^0 = (p, 0) \\ \omega^{\mu\nu} &= (\| \omega_{\mu\nu} \|)^{\mu\nu} \end{aligned} \tag{2.99}$$

The corresponding quantum mechanical formalism in terms of operators $\bar{a} = (\vec{r}, \vec{p})$ and $\bar{H}$ is furnished by Heisenberg equations

$$\dot{\bar{a}}^{\mu} - \frac{1}{i}[\bar{a}^{\mu}, \bar{H}] = 0 \tag{2.100}$$

$$[\bar{a}^{\mu}, \bar{a}^{\nu}] = i\omega^{\mu\nu}; \quad \hbar = 1 \tag{2.101}$$

Newtonian forces are generally nonderivable from a potential. It necessitates a generalization of the Hamiltonian mechanics, first, for local, generally nonpotential forces, and ,second, for nonlocal unrestricted forces.

Birkhoffian mechanics is capable of representing, under sufficient smoothness conditions, all local Newtonian systems. Mathematically speaking, the generalization of the Hamiltonian into Birkhoffian mechanics is characterized by a set of well defined transformations :

(i) From the conventional Hamilton's variational principle in phase space (Hamilton equations) to the most general possible variational principle for first order systems (Birkhoff's equations).

(ii) From the realization of the Lie algebra product in terms of the Poisson brackets to the most general possible realization of the Lie-algebra product in Newtonian Mechanics and

(iii) From the fundamental symplectic structure to the most general possible (but exact) symplectic structure.

A generalization of the Hamiltonian mechanics leads to a generalization of the Heisenberg mechanics.

The $U(SU(2))_{\lambda,\mu}$ mutation is particularly simple as the Pauli matrices verify the following closure properties under the associative product

$$\sigma_i\sigma_j = i\epsilon_{ijk}\sigma_k + \sigma_{ij}1 \tag{2.102}$$

Here, $U(SU(2))_{\lambda,\mu}$ represents the non-associative extension of $SU(2)$ so as to consider the effect of spin mutation. The mutation of the value of the angular momentum, that is, its departure from the value for the exact symmetry, is characterized by the quantity $\lambda+\mu$. Pauli's matrices are also closed under the (λ,μ) product i.e.,

$$\begin{aligned} \sigma_i^*\sigma_j &= \lambda\sigma_i\sigma_j + \mu\sigma_j\sigma_i \\ &= i\frac{\lambda-\mu}{2}\epsilon_{ijk}\sigma_k + i\frac{\lambda+\mu}{2}\delta_{ij}1 \end{aligned} \tag{2.103}$$

It is now clear that the envelope $U(SU(2))_{\lambda,\mu}$ is nonassociative. This can be verified using the property

$$\sigma_1^*(\sigma_2^*(\sigma_3^*\sigma_1)) \neq (\sigma_1^*\sigma_2^*)^*(\sigma_3^*\sigma_1)$$

The value

$$\| \vec{u} \|^2_{A(SU(2))} = \frac{3}{4}, \quad \vec{u} = \frac{i}{2}\vec{\sigma}$$

of the spin under electromagnetic interactions only is now subjected to the mutation

$$\| \vec{u} \|^2 U(SU(2))_{\lambda,\mu} = (\lambda+\mu)\frac{3}{4} \tag{2.104}$$

This can be characterized as the new value of the spin under strong non-self adjoint interactions producing a semiclassical breaking of $SU(2)$.

At the next step Ktorides et al. [Ktorides, 1980] tried to show that the particles considered under mutation are no longer fermions. For this let us consider the mutation of the exponential mapping for the case, of say, σ_3, i.e.,

$$\begin{aligned} E_{\lambda,\mu} &= e^{-\frac{i}{2}\sigma_3}\,|_{U(SU(2))_{\lambda,\mu}} \\ &= \frac{(\lambda-\mu)-1}{\lambda+\mu} + \frac{i}{\lambda+\mu}\sigma_3 \sin[\frac{\theta}{2}(\lambda+\mu)] \end{aligned} \tag{2.105}$$

Now, if we compare it with the conventional form i.e.,

$$\begin{aligned} E &= e^{-\frac{i}{2}\theta\sigma_3}\,|_{A(SU(2))} \\ &= 1\cos(\frac{\theta}{2}) - i\sigma_3 \sin(\frac{\theta}{2}) \end{aligned} \tag{2.106}$$

For $\theta = 2\pi$, $E_{\lambda,\mu} \neq -1$.

Thus, the states which are spinors under electromagnetic interactions are nolonger spinors when subjected to the mutation. For spin $\frac{1}{2}$ nucleons, the explicit realizations can be obtained as

$$\begin{aligned} E_{1,\epsilon} &= e^{-\frac{i}{2}\theta\sigma_3}\,|_{U(SU(2))_{1,\epsilon}} \\ &\cong \epsilon + 1\cos[(1+\epsilon)\frac{\theta}{2}] - i\sigma_3 \sin[(1+\epsilon)\frac{\theta}{2}] \end{aligned} \tag{2.107}$$

where

$$u_i^* u_j = u_i u_j + \epsilon u_i u_j, \quad \epsilon \approx 0$$

$$\| \vec{u} \|^2_{U(SU(2))_{1,\epsilon}} = (1+\epsilon)\frac{3}{4}$$

It is stressed that possible deviation of Pauli principle may be there under strong interaction though very small. But no estimates of this deviation are available from this theory.

Rauch et al. [Rauch 1978] first verified the 4π -periodicity of the spinor wave function using the perfect crystal neutron interferometer. The periodicity factor $\alpha = 715.87 \pm 3.80$ has been observed experimentally. It includes 720^0 within its 2σ error limits. However, more refined experiments are needed to clarify the situation and establish whether the deviation from 4π-symmetry is a real effect. We have studied [Roy 1988] whether there is any deviation from 4π-symmetry due to the fluctuation of the space-time metric in the strong interaction region. In the strong interaction region he metric is assumed to be stochastic in nature. We have considered the probabilistic Finsler space to describe the metric structure of the underlying space-time. For simplicity we have considered the three dimensional probabilistic Finsler space.

The line element in three dimensional Finsler space may be considered as

$$ds^2 = [\delta_{ij} + \alpha\nu_i\nu_j]dx^i dx^j \tag{2.108}$$

where $\alpha << 1$. With no loss of generality we can choose the preferred direction to be the Z-direction. Also let us consider the distance $|\ell_1 - \ell_2| << L$, where L is the typical distance beyond which the directions are completely random. With these considerations, we can write

$$\begin{aligned} < \nu_x(\ell_1)\nu_x(\ell_2) > &= 0 \\ < \nu_y(\ell_1)\nu_y(\ell_2) > &= 0 \\ < \nu_z(\ell_1)\nu_z(\ell_2) > &= 1 \end{aligned} \tag{2.109}$$

We now consider the Hamiltonian H_a of a rigidly charged ball of radius as obtained by Bopp and Haag [Bopp 1950]. We have

$$H_a = 1/2(\vec{p} - e/c\vec{A})^2 + \frac{1}{2\theta}(\vec{M} - 1/c\vec{H})^2 + e\phi \tag{2.110}$$

where, m = Mass of the ball, θ = Moment of inertia of the ball, e = Charge of the ball.

Then, if we calculate the average eigenvalues of the z-component of orbital angular momentum i.e.,

$$< L_z > \psi = (1+\alpha)m\psi = m'\psi \tag{2.111}$$

(say) m' being the effective quantum number. For, $\ell = 1/2, m' = \pm 1/2$, we get,

$$m(1+\alpha) = \pm 1/2; \quad m = \pm\frac{1}{2}(1-\alpha)$$

In this case, α can be taken as 6×10^{-3} [Nielsen 1982]. So,

$$\epsilon \simeq 3 \times 10^{-3}$$

But it falls within a 2σ error limit of the experimental result.

To the author's knowledge, there is no device which could measure the spin directly. It can be measured by measuring the magnetic moment. Then it is possible that the magnetic moment of the neutron will be of anomalous value within the strong interaction region due to the fluctuation of the metric. Thus there might not be any change of fermionic character within the strong interaction region as emphasized by Santilli R.M. [Santilli 1983]. Now if any anomalous value of the magnetic moment of neutron is observed, its origin can be traced back to the stochasticity of the space-time. So this raises new possibility to verify the stochasticity of the Finslerian metric of the micro space-time at least within the range of strong interaction. In a sense, this indicates a deviation from the Riemannian geometry in nature which can be tested experimentally. In that case, the small parameter $\alpha(\sim 10^{-3})$ might indicate the radius of the curvature of the tangent space $T(\nu)$ in the Finslerian manifold.

SPACE-TIME FLUCTUATION AND RANDOM POTENTIAL

"The phenomena of nature are to be referred back to motions of material particles possessing unchangeable moving force, which are dependent upon conditions of space alone"

H. Helmholtz : Über die Erhaltung der Kraft.

3. SPACE-TIME FLUCTUATION AND RANDOM POTENTIAL

3.1. Introduction :

In 1924, Einstein [Einstein 1924] speculated about the fluctuation of space-tim e metric. Menger and his collaborators [Menger 1951] developed the theory of probabilistic metric spaces from the mathematical point of view. In 1983, Schweizer and Sklar [Schweizer 1983] published a monograph collecting mainly mathematical results on probabilistic metric spaces and a few mentions have been made in relation to physical problems. In chapter1 and 2, we have tried to explain the statistical behavior of microparticles considering the fluctuation of space-time metric. We have already mentioned that mathematicians constructed the different types of statistical geometry and studied their probabilistic topology. Among the various types of Random metric spaces, the E-space, as introduced by mathematician Sherwood has much relevance to physical problems. Our model [Roy 1979] of fluctuating metric space is essentially equivalent to that of E-space.The points of an E-space are functions from a probability space (Ω, α, p) into a metric space (M, d). For each pair (p, q) of functions in space, a composite function is defined by

$$d(p,q)(\omega) = d[p(\omega), q(\omega)] \quad \text{for all } \omega \text{ in } \Omega \tag{3.1}$$

In this chapter, we shall discuss the role of the extra variable (which we considered in our framework) in Lagrangian Mechanics both for classical as well as for quantum particles [Roychowdhury 1980, 1982]. This extra variable makes the metric a random one.

3.2. Lagrangian Mechanics :

Almost a century ago, Hertz [Hertz 1956] speculated that the concept of potential could be eliminated from the mechanics. A potential can be generated by proposing unobserved degrees of freedom associated with cyclic coordinates. Indeed, a random potential can be generated from the extra variable in the above mentioned random metric.

In chapter 2, we have already mentioned that if we consider fundamental processes in nature as stochastic processes, then we can get Schrödinger equation or the relativistic wave equation like Klein-Gordon or Dirac equation. Then microscopic characteristics must be averaged at some macroscopic stage so that the probabilistic features can be reinterpreted as the "uncertainties" or "fluctuation" from a physical point of view. On the otherhand, quantum mechanics is usually considered macroscopic with respect to the fundamental stochastic processes in the micro-domain of space-time. Again it is well known that the basic equations of stochastic theory can be formulated for forces which are not derivable from a potential, where as quantum mechanics is incapable of describing such forces. So we shall take up rather heuristically and intuitively following two aspects :

- Fluctuations of space-time
- Reconsideration of Hamiltonian Mechanics

Let us consider the motion of a classical particle of mass m in one dimension described by the position variable x and the directional variable ζ in tangent space attached to x. The metric in an n-dimensional Finslerian manifold may be written as

$$ds^2 = g_{ij}(x,\zeta)dx^i dx^j \tag{3.2}$$

$$i,j = 1,2,\cdots,N$$

For simplicity, let us consider the line element in one dimensional Finslerian manifold as $g(x,\zeta)dx^2$. Assume the special form

$$g(x,\zeta) = g(x) + f(\zeta) \simeq 1 - \alpha g_0(x) + f(\zeta) \quad \text{(say)} \tag{3.3}$$

where, α very is small. When $\alpha = 0$, $g(x)$ simply reduced to the Euclidean metric. If the particle moves freely in this space, Lagrangian can be written as

$$L = \frac{1}{2}m[1 - \alpha g_0(x) + f(\zeta)]\dot{x}^2 \tag{3.4}$$

and the equation of motion can be written as

$$m\ddot{x} + K{g_0}' = 0 \quad \text{where} \quad K = \alpha L\frac{(1-f)}{f'} \tag{3.5}$$

Now, as f is a function of random variable ζ, the above equation may be considered as a ordinary stochastic differential equation with K as a random coefficient. Again, as ${g_0}'$ is a function of x, only by a suitable choice of g_0, it is possible to generate the different types of potential depending upon the distribution function of ζ. ${g_0}'$ and f' denote the derivative with respect to x and ζ respectively.

For example, let us choose $g_0 \simeq x^2$, we get

$$m\ddot{x} + 2Kx = 0 \quad \text{or } m\ddot{x} + \omega^2 x = 0 \tag{3.6}$$

which is the well known equation for random harmonic oscillator. In a subsequent publication [Roychowdhury 1982], we have generalized it to three dimension as well as considering Lagrangian with potential. The metric in a three-dimensional Finslerian manifold may be written as

$$ds^2 = g_{ij}(x,\zeta)dx^i dx^j \tag{3.7}$$

for $i = 1,2,3$ and $j = 1,2,3$. Assuming the general form

$$g_{ij}(x,\zeta) = g_{ij}(x) + f_{ij}(\zeta) \quad = \delta_{ij} + \alpha {g_{ij}}^0(x) + f_{ij}(\zeta) \tag{3.8}$$

for simplicity, let us consider only the diagonal elements. The Lagrangian for a particle of unit mass without potential can be written in polar coordinates (r,θ,ϕ) as :

$$\begin{aligned} L &= 1/2 g_{ij}(x,\zeta)\dot{x}^i\dot{x}^j = 1/2[\dot{r}^2 + \alpha g_{11}^0(r)\dot{r}^2 + f_{11}(\zeta)\dot{r}^2 + r^2\dot{\theta}^2 \\ &\quad + \alpha g_{22}^0(\theta) r^2\dot{\theta}^2 + f_{11}(\zeta) r^2\dot{\theta}^2 + r^2\sin^2\theta\dot{\phi}^2 + \alpha g_{33}^0(\phi) r^2 \sin^2\theta\dot{\phi}^2 \\ &\quad + f_{11}(\zeta) r^2\sin^2\theta\dot{\phi}^2] \end{aligned} \tag{3.9}$$

As a simple case, we consider

$$L = 1/2[1 + \alpha g_{11}^0(r)\dot{r}^2 + f_{11}(\zeta)\dot{r}^2 + r^2\dot{\theta}^2 + r^2\sin^2\theta\dot{\phi}^2] \tag{3.10}$$

For radial motion , we get the equation of motion as

$$\ddot{r} + g_{11}^0{}'/N = 0 \tag{3.11}$$

where

$$N = \frac{\alpha(f_{11} - 1)}{f_{11}'(\zeta)}$$

Since f is a function of random variable ζ , the above equation may be considered as an ordinary stochastic differential equation with N as random coefficient. Again as $g_{11}^{0}{}'$ is a function of r only, we can, by suitable choice of g_{11}^{0}, generate the different types of potential depending on the distribution function of ζ. It may be mentioned that if we start with the Langrangian with a potential, say, harmonic like potential, we get the equation of a random anharmonic oscillator.

Quantum Case :

The Lagrangian for a classical particle of unit mass as mentioned before

$$L = 1/2g_{ij}(x,\zeta)\dot{x}^i\dot{x}^j \tag{3.12}$$

Then the components of generalized momentum should also be defined by

$$p_i = \frac{\partial L}{\partial \dot{x}^i} = g_{ij}(x,\zeta)\dot{x}^j \tag{3.13}$$

Again, as the matrix g^{ij} posses an inverse g_{ij} the kinetic energy T can be defined as

$$T = 1/2g_{ij}p^ip^j \tag{3.14}$$

So, the total energy of the system H is

$$H = 1/2g_{ij}p^ip^j \tag{3.15}$$

Then for quantum particles with unit mass in Minkowski space, the Hamiltonian for radial motion can be written as

$$H = -\frac{\hbar^2}{2}[1 + f^{11}(\zeta r)]\frac{\partial^2}{\partial r^2} \tag{3.16}$$

where $\hbar$ is Planck constant.

In Minkowski space, (not to be confused with four-dimnsional psudo-Euclidean space) the wave function $\psi(r,\zeta_r)$ can be written, at least, locally as

$$\psi(r,\zeta_r) = Q^{1/2}(\zeta_r)\Phi(r)\xi(\zeta_r) \tag{3.17}$$

Then, we get Schrödinger equation with random potential as well as the random coefficient associated with $\frac{\partial\phi}{\partial t}$.

From the above analysis for classical and quantum cases, it is interesting to note that while in classical cases the generation of potential depends on $g(x)$, in quantum case it depends only on $f_{11}(\zeta)$. It is worthmentioning that Aharanov-Bohm hypothesis clearly established the special role of a potential in quantum mechanics. The famous Aharanov-Bohm experiment indicates that, in fact, only the phase factor

$$\exp(\frac{ie}{\hbar c}\oint A_\mu dx_\mu) \tag{3.18}$$

and not the phase is physically meaningful. In otherwords, the phase factor contains more information than the phase. They describe this phase factor as non-integrable i.e., path dependent. The phase factor is the path dependent element of the group $U(1)$, the relevant group related in the quantum mechanical description of atomic phenomena, to the concept of a probability density $\rho(x) = \phi^*(x)\phi(x)$ derived from the complex wave function $\phi(x)$. The structural group $U(1)$ of the bundle in Feynman's treatment, being identical with the associated electromagnetic gauge group $U(1)$ in Mandelstam's treatment, implies the operator of a covariant derivative on the bundle to be given by $D_\mu = \nabla_\mu + ieA_\mu$, A_μ being the electromagnetic potential.

Now in the case of the generalized wave function, $\psi(x,\zeta)$ depends on x and the directional variable ζ, which is a cross-section on fiber bundle constructed over space-time and possesses a gauge or structural group more general than the group $U(1)$. But if we take

$$\psi^*(x,\zeta)\psi(x,\zeta) = |Q^{1/2}(\zeta)|^2|\phi(x)|^2|\chi(\zeta)|^2 = \rho(x) \tag{3.19}$$

for $|Q^{1/2}(\zeta)\chi(\zeta)|^2 = 1$, which means $|S(\zeta)|^2 = 1$, then $U(1)$ will be the relevant group related in quantum mechanics. In this case, $\frac{\partial S}{\partial t} = 0$, so that the potential

$$V = (1 - f^{11})(1/S)\frac{\partial S}{\partial t} \tag{3.20}$$

vanishes. It indicates that if ψ is a globally defined function, then there will be no potential arising out of randomness of space-time. But if we take the probabilistic geometry as a more fundamental concept than the concept of ψ function defined over Hilbert space, then there will be a non-vanishing potential due to the fluctuation of the metric. But the ψ-function can not be considered as the globally defined function over the whole space-time.

The above idea of generation of potential due to the fluctuation of space-time metric raises immediately some interesting questions regarding the operational aspects of the measurement; whenever one is going to measure in micro-domain one has to adopt a particular type of averaging procedure so as to generate a different potential or to perturb the potential in a particular way. As a result, even in the classical case, whenever one performs a measurement in the micro-domain, the system will also be perturbed according to our hypothesis. As far it is known that in quantum mechanical framework, it is possible to attach some meaning to this sort of perturbation as "Uncertainty" through Heisenberg principle. Here,we are having a different kind of uncertainty other than quantum mechanical one. For example, it would be the effect of measurement of even one of the conjugate pair of observable field components. This new indeterminacy would exist in classical as well as in quantum field theory. In fact, Pauli [Pauli 1923] also pointed out that the connection of the mathematical limitation of QED with Gravitation as mentioned by Landau and Klein [Landau 1962] i.e., $p \sim \frac{\hbar}{\ell_0}$ (p, being the momentum, $\ell_0 = \sqrt{\chi \hbar c}$, χ is Einstein gravitational constant) seems to hint at the indeterminacy in space-time of light which is governed by probability laws in quantized field theory, invariant with respect to wider group of general relativity.

3.3. Lamb Shift and Hydrogen Atom :

It is evident from the above analysis that since the fluctuation of space-time metric perturbs the potential present in the Lagrangian, we have calculated the perturbation of energy levels of the Hydrogen atom in the non-relativistic framework similar to original derivation of Bethe [Bethe 1947]. A shift in the energy level of the Hydrogen atom [Roychowdhury 1987] has been found due to this perturbation . This is similar to Lamb shift. For simplicity, the Lagrangian is taken as

$$\mathcal{L} = \frac{1}{2}[1 + \alpha_0 g^{11}(r) + f^{11}(\zeta_r)]p_r{}^2 \tag{3.21}$$

where, $[1 + \alpha_0 g^{11}(r) + f^{11}(\zeta_r)]p_r$ is the radial part of generalized momentum, assuming $f^{22}(\zeta) = 0$, $f^{33}(\zeta) = 0$. The corresponding Hamiltonian for quantum particle can be written as

$$H = -\frac{1}{2}\hbar[1 + f^{11}(\zeta_r)]\frac{\partial^2}{\partial r^2} \tag{3.22}$$

Then the resulting Schrödinger equation in a potential $V(r)$ becomes

$$-\frac{\hbar^2}{2\mu}\nabla^2\phi + (1 - f_{11})V(r)\phi = (1 - f_{11})E\phi \tag{3.23}$$

For Coulomb potential $V(r) = -\frac{Ze^2}{r}$ we can solve the above stochastic differential equation for energy eigenvalues of Hydrogen atom.

In spherical coordinates

$$\Phi_{nml}(r,\theta,\phi) = R_{nl}Y_{lm}(\theta)Z_m(\phi) \tag{3.24}$$

the radial part of the equation

$$L\Phi = \lambda M\Phi \tag{3.25}$$

where $L = \frac{d^2}{dr^2} - V$, $\lambda = (E-C)\frac{2\mu}{\hbar^2}$, $M = 1 - f_{11}$, μ is the reduced mass.

If $f_{11} = 0$, then the equation reduces to usual Schrödinger equation with Coulomb potential. We assume that L and M can be expressed in the following way

$$\begin{aligned} L &= L_0 + \epsilon L_1 \\ M &= M_0 + \epsilon M_1 \\ M_0 &= 1, \quad L = \nabla^2 - V(r) \end{aligned} \tag{3.26}$$

For a particular case

$$\langle f_{11} \rangle = \epsilon_1(\beta e^{-\alpha r}) \tag{3.27}$$

ϵ_1 is a small parameter and α, β are constants. Now we know

$$\lambda_{10} = (L_1, \Phi, \phi), \lambda_{01} = -\lambda_{00}(M\Phi, \phi)$$

λ_{00} is the eigenvalue for a deterministic Coulomb potential. Since $\lambda_{00} \sim \frac{1}{n^2}$ for large n, λ_{01} may be neglected, hence we shall calculate λ_{10} only. After some calculations, we obtain λ_{10} as

$$\lambda_{10} \sim (\frac{\beta Ze^2}{4n^3})\alpha e^{-\frac{1}{2}p^2} I_3(-\frac{1}{2}p^2) \tag{3.28}$$

with $l = 1$ for large n and $p^2 = \frac{na_0}{8Z(n+1)}$.

Similar results hold for any l. For $l > 1$, the exponential term falls quickly $\sim \frac{1}{l}$. It is important to note that the Lamb shift is also negligible for the $2p$

state compared with the S-state. The above result yields a shift of energy levels comparable to that of the Lamb shift. For the general case i.e.,

$$\langle f_{11}\rangle = \phi(r); \quad \lambda_{10} \sim \frac{\beta Z e^2}{2a_0}.\frac{1}{3}f(\frac{na_0}{2Z})$$

$$f = \int_0^{\infty} \phi((8\frac{na_0}{2Z})\epsilon^2)[J_1(t)]^2 t dt \tag{3.29}$$

The only difference is that the numerical value of the shift will be different depending on the parameters β, ϵ, α. It should be mentioned that if we take the distribution such that $f_{11} \sim e^{-\alpha r}$, then the Coulomb potential will be modified as

$$\frac{e}{r}[1 - e^{-\alpha r}]$$

3.4. Fifth Force :

Over the past one decade, evidence has been accumulated in favor of deviation from Newtonian inverse square law of actions of the gravitational interaction. It has been in the literature for quite some time [Stacy 1987] that any deviation from Newtonian's law of macroscopic range is to be expected to take the form of a general potential

$$V = -\frac{G_{\infty}m}{r}\left[1 + \sum_{k=1}^{n} \alpha_k e^{-\frac{r}{\lambda_k}}\right] \tag{3.30}$$

where, G_{∞} is the Newtonian gravitational constant at large distances, m the mass and r the distance at which the potential is measured, α_k is the strength of the k-th Yukawa potential with range λ_k. This general parametrization for the non-Newtonian potential of the above form has been tested by several groups of experimentalists using totally different techniques, over the past few years [Fishcbach 1986; Speake 1988].

The general outcome of these experiments even on retaining only one term $k = 1$, the data can be fitted only for a non-zero value of $|\alpha_1| \sim 10^{-2}$ and $\lambda_1 \sim 10^2$ to 10^3 meters. Donald et al. [Donald 1988] even find that the above potential with two terms [$k = 1$ and 2] can also account for the experimental data provided α_1 and α_2 be of the opposite signs.

The final conclusion of the present experimental findings seem to be that, though the numerous analysis of laboratory experiments and tracking data of natural bodies and of artificial satellites and spacecrafts in the solar system have shown "null" results as far as detection of deviation from

Newtonian inverse square law of gravitational interaction is concerned, the "geophysical window" between the laboratory and the astronomical null results has revealed incontrovertible evidence for the existence of a Yukawa type deviation with an intermediate range of about a few hundred meters. With just one Yukawa term in the potential,

$$V(r) = -G_\infty \frac{m}{r}(1 + \alpha e^{-\frac{r}{\lambda}}) \tag{3.31}$$

the geophysical data can be accounted for quantitatively if α and λ have the values

$$|\alpha| \sim 10^{-2} \quad \text{and} \quad \lambda \sim 10^2 \quad \text{to} \quad 10^3 \quad \text{meters.}$$

All the experiments so far indicate the values of α and λ in the range mentioned, though there is yet no consequences as to whether the deviation from Newtonian gravity is attractive or repulsive. The general indication is that a repulsive Yukawa term is the more favored form. One significant feature of these experimental result is that the non-Newtonian deviation is intimately related to the matter aggregation. This is perhaps of terrestrial origin as experiments on the measurement of the Newtonian gravitational constant conducted for the above terrestrial surface have indicated null deviation from the hitherto accepted universal value.

Fuji et al. [Fuji 1988] suggested that this deviation may also arise from a new interaction which couples to nuclear Isospin and Baryon number. Though the data at present does not seems to null out the possibility of composition dependence of the "fifth force", the accuracy with which such dependence can be established at present is a moot point. The evidence however is certainly in favor of a composition independent short range force with a single Yukawa repulsive force. With the above experimental status of the "fifth force" in mind, we have suggested in a recent paper [Roy 1988] that it may be possible to understand the origin of this fundamental phenomenon in a more general macroscopic framework without going into the details of the sub-atomic processes that may be involved in the generation of the "fifth force".

At first we shall discuss the essentials of stochastic differential geometry on Finslerian manifold which we use to describe the phenomenon of "fifth force" and its geophysical origin.

Stochastic Differential Geometry :

Consider a set of element M, ordered with the aid of an attribute ξ and coordinate x for each element. Assuming that the mapping of attributes ξ

on the coordinates X containing a random valued parameter such that

$$x = X(\xi, \zeta) \tag{3.32}$$

we assume that the distribution of possible values of ζ are given by a normalized definite density $\rho(\zeta)$

$$\rho(\zeta) \geq 0, \quad \int_{-\infty}^{+\infty} \rho(\zeta) d\zeta = 1 \tag{3.33}$$

Therefore, the mean value over the region of random parameters ζ, which will be denoted by the usual angular brackets, can be determined from the equation :

$$\bar{X} = \langle X \rangle = \langle X(\xi \mid \zeta) \rangle = \int X(\xi \mid \zeta) \rho(\zeta) d\zeta \tag{3.34}$$

in which the space $R_4(x)$ as defined above is stochastic space. It is further assumed that it is differentiable and

$$\text{Det } \left(\frac{\partial X_\alpha}{\partial \xi_\beta}\right) \neq 0 \tag{3.35}$$

with α and β running over the dimension of the space under consideration. It is now possible to construct the following metric in $R_4(x)$ as

$$ds^2 = G_{\mu\nu}(x, \zeta) dx^\mu dx^\nu \tag{3.36}$$

with

$$G_{\mu\nu}(x, \zeta) = g_{\alpha\beta}(\xi) \left(\frac{\partial \xi^\alpha}{\partial x_\mu}\right) \left(\frac{\partial \xi^\beta}{\partial x_\nu}\right) \tag{3.37}$$

as a tensor. If the size of the fluctuation is small then we may write

$$G_{\mu\nu}(x, \zeta) = G_{\mu\nu}(\bar{x}) + \alpha f_{\mu\nu}(\zeta) \tag{3.38}$$

where, α is a numerical parameter $\alpha < 1$ and $G_{\mu\nu}(\bar{x})$ is the background stable metric free from fluctuation and $f_{\mu\nu}(\zeta)$ is the fluctuating part of the metric.

Limit On The Value Of α :

Blokhintsev [Blokhintsev 1975] estimated the magnitude of the parameter α on very general arguments. We shall consider his arguments in details in chapter 4. He found a relation as

$$\alpha = \frac{{\Lambda_g}^2}{\Lambda^2} \tag{3.39}$$

where, $\Lambda = (\frac{8\pi G\hbar}{c^3})^{\frac{1}{2}} \approx .82 \times 10^{-32}$ cm. Therefore, α will be small as long as $\Lambda >> \lambda_g$. Taking the value of Planck length

$$\Lambda_g \simeq 1.6 \times 10^{-33} \text{ cm}; \quad \alpha \sim 10^{-2} \quad \text{to} \quad 10^{-3} \tag{3.40}$$

In the domain of elementary particle physics, several authors [Nielson 1982, 1983] considered the deviation of the Lorentz invariant flat metric by a small parameter. That small parameter has been estimated from $K_0 - \bar{K}_0$ decay also and found to be of the order of 10^{-2} to 10^{-3}.

Non-Newtonian Deviation :

Let V_N denotes the Newtonian gravitational potential. Then using the effect of metric decomposition as

$$g_{\mu\nu}(x,\zeta) = \bar{g}_{\mu\nu}(x) + \alpha f_{\mu\nu}(\zeta) \tag{3.41}$$

we get,

$$V_{\text{eff}} = V_N(1 + \alpha\langle f_{00}(\zeta)\rangle)_{\zeta=r} \tag{3.42}$$

In order to calculate the average of the fluctuations, we have to introduce a plausible density distribution function for the random variable ζ. A simple Poisson distribution is a natural choice as it has been proved fruitful in a number of diverse application of stochastic geometry. Using Poissonian density distribution $\rho(\zeta)$, we get

$$V_{\text{eff}} = V_N(1 + \frac{\alpha r}{\mu} e^{-\mu r}) \tag{3.43}$$

with $\mu = \alpha + \beta$; $\quad f_{00}(\zeta) = e^{-\beta\zeta}$ and $\rho(\zeta) = N^{-r\zeta}$

Therefore, μ^{-1} would be of the range of the non-Newtonian deviation. Taking $\mu \sim r \sim \beta$, we get

$$V_{eff} = V_N(1 + \alpha e^{-\mu r}) \tag{3.44}$$

Again,

$$|\frac{\delta\rho(\zeta)}{\rho(\zeta)}| = r|\delta\zeta| \tag{3.45}$$

with $r = \frac{1}{|\delta\zeta|}|\frac{\delta\rho(\zeta)}{\rho(\zeta)}|$.

If we relate density fluctuation in terrestrial matter with the quantity $|\frac{\delta\rho(\zeta)}{\rho(\zeta)}|$ which gives the magnitude of the fluctuation, then

$$|\frac{\delta\rho(\zeta)}{\rho(\zeta)}| \sim |\frac{\delta\rho_{\text{matter}}}{\rho_{\text{matter}}}| \sim |\frac{\Lambda_g\Lambda}{3}| \sim |\alpha|^{\frac{3}{2}} \tag{3.46}$$

Using the value of $\alpha \sim 10^{-2}$ to 10^{-3}, we have

$$|\frac{\delta\rho(\zeta)}{\rho(\zeta)}| \sim 10^{-3} \quad \text{to} \quad 10^{-4.5} \tag{3.47}$$

If we further assume that the fluctuation in matter density averaged out in a matter thickness greater than $|\delta\zeta| \sim 10^{-3}\text{m}$ to 10^{-2}m which is consistent with present day Geophysical experiment [Stacy 1987], then,

$$\mu \sim r \sim 1m^{-1} \quad \text{to} \quad 10^{-2.5}m.$$

This gives a range for the non-Newtonian force of the order of 1 meter to a few hundred meters. The present experimental evidence favors the higher value and we should expect that only matter density fluctuations from a shell of matter of thickness $\sim 10^{-2}$ nearest to the point of measurement contributes to the non-Newtonian Gravity.

3.5. Origin of Quantum Potential :

We have already pointed out in chapter 2 that quantum potential plays a significant role in Bohm-Vigier hidden variable theory. Vigier and his collaborators [Vigier 1984] raised a lot of interest about the role of quantum potential in explaining the mysterious nature of quantum statistics. However, all these new proposals still remain mysterious unless one can understand the origin of the quantum potential in terms of subquantal motion.

We have already shown that the fluctuation of space-time metric generates potential in three dimensional cases. But those are classical potentials. Here we shall show that the quantum potential can also be traced back to its origin to fluctuation of space-time metric. Takano [Takano 1961] introduced the concept of fluctuation of metric by assuming that if the elementary particle exists, space-time is distributed or fluctuated and the metric can be written in the form

$$\tau'^2 = \tau^2 - a^2, \quad \tau^2 = x_\mu{}^2 = x_0^2 - x_1{}^2 - x_2{}^2 - x_3{}^2 \tag{3.48}$$

with having the probability distribution $P(a)$. In this picture, the modification of the metric fundamental form may be understood geometrically as the introduction of the fifth dimension, which is conditioned by the probability distribution. This causes the fluctuating four-dimensional space structurally to be a five-dimensional probability distribution of four-dimensional subspaces corresponding to various values of a. The physical quantity in this theory can be defined as a weighted average of the corresponding quantity in four-dimensional subspace.

Let us consider $F^4(x,\zeta)$ as the four-dimensional Finsler space when $x \in R^4, \zeta \in T(R^4)$ and $M^8(\xi)$ as eight dimensional space. Then $F^4(x,\zeta) \subset M^8(\xi)$. So,

$$ds^2 = g_{\mu\nu}(\xi)dx^\mu dx^\nu \tag{3.49}$$

for,

$\mu = 0, 1, 2, 3, \cdots, 7; \quad \nu = 0, 1, 2, 3, \cdots, 7$
$x^i = (x, ict), \quad i = 1, 2, 3, 0; \quad x^j = (\vec{\zeta}, \zeta_0) \quad$ for $\; j = 4, 5, 6, 7$

Now if we consider that space is preferred with respect to the fourth component of ζ, it reduces to the form taken by Takano. Here, ζ is taken as a random vector with $p(\zeta)$ as the probability distribution. Again this ζ can be identified with the velocity vector i.e., M^8 can be identified with the eight-dimensional relativistic phase-space. So,

$$\tau'^2 = \tau^2 - \zeta^2 = |\,\vec{x}\,|^2 - (c^2 + \zeta_0{}^2)^2 t^2 \tag{3.50}$$

Then it can be thought of as due to the fluctuation of c itself. The distribution function is normalized as

$$\int_{-\infty}^{+\infty} p(\zeta)d\zeta = 1$$

This whole scheme can be fitted [Roy, 1986] into the pattern of a five dimensional Kaluza-Klein theory i.e.,

$$ds'^2 = a_{mn}dx^m dx^n = a_{ij}dx^i dx^j + a_{55}dx^5 dx^5 \tag{3.51}$$

$m, n = 1, 2, 3, 4, 5$
$i, j = 1, 2, 3, 4$
$a_{ij} = \text{diag}(+1, +1, +1, -1)$

Here, a_{55} is a random component of the metric tensor a_{mn} in contrast to the usual Kaluza-Klein framework where a_{mn} is a non-random tensor. By suitable choice of the coordinate system, the metric tensor in five dimension can be shown to be independent of the fifth coordinate i.e., $a_{mn,5} = 0$ which implies

$$\bar{g}_{55} = < a_{55} > = \phi(x) \tag{3.52}$$

Then the Einstein equation in five-dimension can be written as

$$R_{mn} - \frac{1}{2}g_{mn}R = -KT_{mn} \tag{3.53}$$

The Ricci tensor may be computed as

$$R_{55} - \frac{1}{2} g_{55} R = -\frac{1}{2}\ \Box\phi \tag{3.54}$$

and thus

$$R_{55} = -8\pi\rho {u_5}^2 \tag{3.55}$$

ρ being the mass density in five dimensions and u_5 the fifth component of the velocities in five dimensions.

This ρ is related to the density according to

$$\rho\left(1 - \frac{u_5^2}{\phi(x)}\right) = \rho_0 \tag{3.56}$$

So,

$$-\frac{1}{2}\Box\phi = -8\pi\rho u_5^2 = -8\pi(\rho - \rho_0)\phi(x) \tag{3.57}$$

and,

$$\begin{aligned} ds^2 &= d\tau^2[1 - \frac{u_5^2}{\phi}\,\frac{h^2}{2m^2c^2}] \\ &= d\tau^2[1 + \frac{\Box\phi}{\phi}.\frac{h^2}{m^2c^2}.\frac{1}{D}] \end{aligned} \tag{3.58}$$

where, D is a constant, $\hbar$ is Planck constant, m is the mass of the particle under considerations. The above expression seems to be very interesting from the viewpoint of de Broglie's theory of Double solution. If we assume that the space-time is Euclidean we can put

$$\begin{aligned} ds^2 = (\tfrac{M_0}{m_0})^2 d\tau^2 &= (\frac{M_0}{m_0})^2 G^0_{ij} dx^i dx^j \\ &= r_{ij} dx^i dx^j \end{aligned} \tag{3.59}$$

So,

$$r_{ij} = (\frac{M_0}{m_0})^2 G^{(0)}_{ij} = [1 + \frac{\Box\phi}{\phi}(\frac{h^2}{m^2c^2})]\delta_{ij} \tag{3.60}$$

M_0 denoting the variable rest mass, i.e.,

$$M_0 = \left[m_0^2 + \frac{h^2}{c^2}\frac{\Box\phi}{\phi}\right]^{\frac{1}{2}} \tag{3.61}$$

It should be noted that this quantum potential is valid only for spin-zero particles.

Subsequently Ranganathan [Ranganathan 1988] generalized this approach to spin-1/2 case and constructed the spin dependent quantum potential. Here, it is shown that if space-time is endowed with scale covariant Finsler structure, then part of the action due to the intrinsic Weyl field can be reinterpreted as the Dirac particle and its fluctuations the quantum potential. Let us try to explain this in the following way .

Consider a scalar field Ω on a Weyl manifold. Then under dilations

$$\begin{aligned} g_{\beta\gamma} \to \bar{g}_{\beta\gamma} &= \exp[2\omega] g_{\beta\gamma} \\ \Omega \to \bar{\Omega} &= \exp[-\omega]\Omega \\ R \to \bar{R} &= \exp[-2\omega][R - 6(\Phi^{\alpha}_{,\alpha} + \Phi^{\alpha}\Phi_{\alpha})] \end{aligned} \tag{3.62}$$

Then the corresponding scale invariant action is

$$\begin{aligned} I &= \int d^4x\sqrt{-g}\Omega^2[R - 6(\Phi^{\alpha}_{,\alpha} + \Phi^{\alpha}\Phi_{\alpha})] \\ &+ a\Omega^4 + b\Omega^{\alpha}\Omega_{\alpha} - 12b\Omega\Omega^{\alpha}\Phi_{\alpha} \end{aligned} \tag{3.63}$$

where a and b are arbitrary constants and the last term occurs due to the Finslerian nature of the space. The Weyl field can be splitted into a scale invariant and a scale covariant form as

$$\Phi = \Phi_c + \Phi_I \tag{3.64}$$

Again it has been shown [Ranganathan, 1988] that the invariant ppart may be written as a Dirac current by considering its covariant derivative term. When we project on to the local tetrad, the derivative in terms of spinor components is

$$\delta_{\mu} = V^{\alpha}_{\mu}\delta_{\alpha} = V^{\alpha}_{\mu}\sigma^{AB}\delta_{AB} \tag{3.65}$$

as we have two Weyl fields of opposite handedness, the derivative operators acting on them must have opposite sign (due to the opposite orientation of the basis when projected back from spinor to tetrad consistently. Thus we see that instead of vanishing as a total derivative from the action, we have

$$\Phi^{\alpha}_{I;\alpha} = \bar{\Psi}\overline{\gamma^{\alpha}\delta_{\alpha}}\Psi \tag{3.66}$$

We now use the fact that the spinors are Grassmann valued and dilation invariant to write

$$\{\Psi(x + dx, t)\Psi(x, t)\} = K\Omega\delta^3(dx) \tag{3.67}$$

Here, braces denote an anticommutatior and K is a constant. It should be noted that this is a classical result derived on the basis of consistency of the Grassmann and scale invariant structures under translation and not a quantum mechanical postulate. Now consider the term quadratic in Φ_I^α. This can be written as

$$\begin{aligned}\int d^4x\sqrt{-g}\Phi_I^\alpha\Phi_{I\alpha} &= \int\int d^4x d^3(dx)\rho(dx)\sqrt{-g} \\ &\times \ [\Phi_I^\alpha(x+dx,t)\Phi_{I\alpha}(x,t)]\end{aligned} \tag{3.68}$$

where $\rho(dx)$ is an appropriate normalized smearing function.

Then we have

$$\int d^4x\sqrt{-g}\Phi_I^\alpha\Phi_{I\alpha} = \int d^4x\sqrt{-g}K\Omega\Psi\Psi \tag{3.69}$$

and the action is

$$\begin{aligned}I &= \frac{\int d^4x\sqrt{-g}\Omega^2[R-6(\bar{\Psi}}{\gamma^\alpha\delta_\alpha\Psi+K\Omega\Psi\Psi} \\ &+ \ 2\bar{\Psi}\gamma^\alpha\Psi_{c\alpha}\Psi+\Phi_c^\alpha\Phi_{c\alpha} \\ &+ \ 2b\Phi_c^\alpha\frac{\Omega_\alpha}{\Omega}]+a\Omega^4+b\Omega^\alpha\Omega_\alpha\end{aligned} \tag{3.70}$$

By varying Ψ and $\bar{\Psi}$ we obtain

$$(\Omega^2\bar{\Psi}\gamma^\alpha\Psi)_{;\alpha} = 0 \tag{3.71}$$

Using this equation and by varying Φ_c^α we can have

$$\Phi_c^\alpha = -b\frac{\Omega^\alpha}{\Omega} \tag{3.72}$$

Thus the covariantly varying part of Weyl's field is not a dynamical field at all and may be removed from the action. The scalar field couples in precisely the right way for the Higgs mechanism to occur and in its minimal energy state (neglecting the fermionic coupling contribution) is

$$\Omega = \sqrt{-\frac{R}{a}} \tag{3.73}$$

The action is the Einstein-Dirac one with the Dirac particle having a mass $K\sqrt{-R/a}$ and subject to a quantum potential Ω_a/Ω. The action can be written as

$$I = \frac{\int d^4x\sqrt{-g}\Omega^2\{R-6[\bar{\Psi}}{\gamma^\alpha\delta_\alpha\psi + K\Omega\bar{\Psi}\Psi} - 2b\bar{\Psi}\gamma^\alpha(\frac{\Omega_\alpha}{\Omega})\ Psi]\} + (6b^2+b)\Omega^\alpha\Omega_\alpha + a\Omega^4 \tag{3.74}$$

It is now clear from the above analysis that if the additional field of a Finsler space is interpreted as a Dirac field rather than the electromagnetic field as Weyl proposed, the theory is consistent.

De Broglie in his theory of double solution emphasized that even if the particle is not subjected to any gravitational or electromagnetic field, its possible trajectories are the same as if space-time possess the non-Euclidean metrics defined by r_{ik}. In the theory of double solution the subquantal vacuum plays a crucial role in considering the quantum potential. But so far no explicit formulation of quantum potential exists in the theory of Double solutions. In our framework [Roy, 1986] of stochastic space-time, if the particle is considered in probabilistic non-Riemannian geometry in the microdomain, the quantum potential for both spin-0 and spin-1/2 particles can be related to the fluctuations of the metric.

3.6 Quantum Statistics and Random Trajectories :

The quantum potential model of Bohm [Bohm 1952] has been shown to reproduce typical quantum configurations such as the double-slit experiment in terms of well defined space-time trajectories which do not coincide with the classical areas since the quantum potential reflects the experimental context as a whole. Furthermore recent experiments on neutron self interference [Badurek 1983] have been equally reproduced by a spin generalized version of this approach [Dewdney 1986] in terms of well defined trajectories and continuously variable spin vectors, implying again the distinguishable character of the quantum elements. The quantum statistics "paradox" of indistinguishability was then resolved by showing that Bose-Einstein and Fermi-Dirac statisitcs correspond to causal, space-time behavior of distinguishable particles correlated by action at a distance [Kyprianidis 1984]. This is based on the idea of Tersoff and Bayer [Tersoff 1983]. It should be mentioned that Kunte [Kunte 1977] got this result long before Tersoff and Bayer.

Tersoff et al claimed that even distinguishable particles can obey Bose-Einstein or Fermi-Dirac statisitcs contrary to the indistinguishable nature of quantum particles. Their work suggests that a reconciliation of quantum mechanics with the classical idea of distinguishablity might be possible by reassesing the traditional assumptions. It can be shown [Dasgupta 1990] that neither is the idea of distinguishability reconciled with quantum statisitcs nor is there any necessity of reassesment of traditional probability assumptions as explained by them.Within the traditional set up consider three marbles (particles a, b,c) which are randomly distributed with equal probability over three cells(regions in the state space , to be dnoted by bars). If the marbles are distinguishable then there are $3^3 = 27$ possibilities , viz.

$$| \, abc \, | - |, | - | \, abc \, | - |, \cdots, | \, b \, | \, a \, | \, c \, |, | \, a \, | \, b \, | \, c \, |,$$

each with probability $\frac{1}{27}$. But if the marbles are indistinguishable then denoting a marble by an asterisk there are only 10 possibilities, viz.

$$| * * * | - | - |, | ** | * | - |, | ** | - | * |, | * | * | * |, \cdots, | - | - | * * * | .$$

Some micro particles follow the second scheme of indistinguishable marbles with uniform associated propbabilities for each distinguishable arrangement, i.e. $\frac{1}{10}$ here. Therefore they are indistinguishable as the associated probabilities are not compatible with those of distinguishable particles. The Fermi-Dirac (FD) particles obey the additional restriction that a single state is occupied by at most one particle.

It may be mentioned that if $P(ab) = P(ba)$, where ab represents marble a in the first position and b in the second position, then some authors like to define marbles [Bach 1985] a and b as "indistinguishabl". But if the particles are indistinguishable then ab and ba represent the same state and as such associated probabilities seperately to ab and ba becomes meaningless.In traditional theory the fact that two events have the same probability does not mean that these events are identical or indistinguishable. Kunte[Kunte 1977] gave a characterization of BE statistics as a compound distribution of a multinomial distribution with Dirichlet distribution. This characterization gives us a method of putting one marble at a time in cells in such a way that BE statistics is obtained.

Let $\vec{W} = (W_1, W_2, \cdots, W_{k-1})$ be a random vector uniformly distributed in the region

$$\Delta = \{W_1, W_2, \cdots, W_{k-1} : W_i \geq 0 \text{ and} \sum_{i=1}^{k-1} W_i \leq 1\} \qquad (3.75)$$

Therefore, the density of W, $f_w(\vec{W}) = c$ on Δ c is such that the total probability mass of W when integrated over Δ 1. Consider the following Dirichlet integral :

$$\int_\Delta \cdots \int W_1^{n_1-1} \cdots W_{k-1}^{n_k-1-1}(1 - W_1 - \cdots - W_{k-1})^{n_k-1} d\vec{W} =$$

$$\frac{\Gamma(n_1)\cdots\Gamma(n_k)}{\Gamma(n_1 + n_2 + \cdots + n_k)} \tag{3.76}$$

for any $n_1 .... n_k$. This type of multiple integral can be reduced successively to a single integral [Whittaker 1958]. To compute c, let $n_1 = ... = n_k = 1$ in formula(3.76). Since the total probability mass should be 1, for uniform distribution over ∇

$$\begin{aligned} fw(\vec{W}) &= (k-1)!, \quad \text{if} W_i \geq 0, \ \sum_{i=1}^{k-1} W_i \leq 1 \\ &= 0 \quad \text{otherwise} \end{aligned} \tag{3.77}$$

Incidentally the distribution

$$\begin{aligned} g_w \vec{W}) &= \frac{\Gamma(n_1 + \cdots + n_k)}{\Gamma(n_1)\cdots\Gamma(n_k)} W_1^{n_1-1} \cdots W_k^{n_k-1} \\ &\times \ (1 - W_1 - \cdots - W_{k-1})^{n_k-1} \end{aligned} \tag{3.78}$$

on Δ called the standard Dirichlet distribution [Johnson 1972]. This is a multivariate generalisation of the beta distribution and can be obtained as the ratio of a set of chi-variables to the sum total of several independent chi-variables(positive square root of a chi-square variable). The density in (3.77) is a particular case of the above with $n_i = ... = n_k = 1$. Let $\vec{N} = (N_1, N_2, ...N_k$ be a non-negative integral valued random vector such that given $\vec{W} = \vec{w}$, N has a multinomial distribution with parameters $n, w_1, w_2, ...w_k$ where

$$W_k = 1 - \sum_{i=1}^{k-1} W_i$$

Thus

$$\begin{aligned} P(N_1 &= n_1, \cdots, N_k = n_k \mid \vec{W} = \vec{w}, \sum_{i=1}^{k} n_i = n) \\ &= \frac{n!}{n_1! n_2! \cdots n_k!} w_1^{n_1} w_2^{n_2} \cdots w_k^{n_k} \end{aligned} \tag{3.79}$$

This multinomial distribution when integrated over Δ the uniform density (3.77) gives BE statisitcs. With an application of the formula (3.77)

$$\int_{\Delta} P(\vec{N} = n \mid \vec{W} = \vec{w}) f_{\vec{w}}(\vec{W}) d\vec{W} = \binom{n+k-1}{k}^{-1};$$

where $\sum_{i=1}^{k} n_i = n$.

In Maxwell-Boltzmann(MB) statisitcs, equal probability $\frac{1}{k}$ for a marble to go into any cell has been assumed whereas in BE statistics we have to assume equal expected probbability with the uniform a priori distribution $E(w_i) = \frac{1}{k}$. Let $\vec{W}$ be a random vector uniformly distributed in the region Δ. The apriori probability distribution of a marble going into a cell remains uniform for every cell and every marble. All possible probabilities of distinguishable particles are averaged out. As a result the distinguishable arrangements of particles become prominent but the particles loose their distinguishability because of averaging. Tersoff and Bayer considered the same apriori distribution as in (3.75) but used the term arbitrary weighting which is different in concept than that of random uniform weighting. Here lies the misunderstanding. Infact Feller [Feller 1957] mentioned that the possibility of choosing the cell at random without regard to its content even within the framework of traditional probability theory resulting in indistinguishability of cells.

Regarding the claim in Tersoff et al work that distinguishable particles can obey BE/FD statisitcs, we feel that it is not appropriate to attribute the characterisitc of the averaged probabilities of the particles to the original distinguishable particles. It is not that the distinguishable particles suddenly decide to bahve in an indistinguishable manner but the fact is that their original characteristic of distinguishability is totally lost after averaging. The heigh of a particular individual can not be recovered from the mean heght of a group of persons to which he belongs. From the Bayesian point of view one may say that tha particles loose their distingushability after being averaged by the uniform apriori istribution on Δ .The mixing density in(3.77) can be interpreted as the a priori density of the random variable $\vec{W}$.

This type of representation via multinomial distribution is not possible for FD statisitcs because of the following reason. For a given

$$w = (w_1, w_2, \cdots w_k), \sum_{i=1}^{k} = 1$$

the multinomial distribution allows more than one particle per cell. Now these probabilities when averaged over any positive density of $\vec{W}$ give a positive probability for more than one particle per cell which is clearly not compatible with the characteristic FD particles. Now if we consider a slightly differs a priori distribution, we get BE statistics as a limiting case. for example, let

$$f_{\vec{w}}(\vec{W}) = (k-1)!(1+\frac{w_1 - w_2}{Lk}) \tag{3.80}$$

ON δ, $\int_\Delta f(\vec{W})d\vec{W} = 1$, $L \geq 1$ is a constant, and this a priori distribution being compounded with the multinomial gives

$$P(N = n_1, n_2, \cdots N_k = n_k) = \frac{n!(k-1)!}{(n+k-1)!}(1+\frac{n_1 - n_2}{(n+k)Lk}) \tag{3.81}$$

Then in the limiting case as $k \to \infty$or $L \to \infty$ we get back BE statistics. $L \to \infty$ implies that the peerturbation over a uniform a priori distribution is negligible. We note that in the second part of (3.80) there is an odd function of w and some other odd function $g(\vec{W})$ of $\vec{W}$ can also be considered. This can loosely be interpreted as follows. On certain restricted sets of Δ, the restriction being on the first two coordinates, the marbles are indistinguishable. One may of course take the perturbation of w in some other form, say a multiple of W_1W_2 with suitable adjustment of the constant $(k-1)!$ therein.

$$f_{\vec{w}}(\vec{W}) = a(1 + w_1 w_2) \quad \text{on} \quad w_i \geq 0, \sum_{i=1}^{k-1} w_i \leq 1 \tag{3.82}$$

Where a can be expressed in sych a way that we can write

$$\int_\Delta f d\vec{W} = 1\,.$$

So,

$$\begin{aligned} p &= \frac{(k+1)!}{k^2+k+1}\int_\Delta (1+w_1w_2)\frac{n!}{n_1!\cdots n_k!} \\ &\times \ w_1^{n_1}\cdots w_k^{n_k}dw_1dw_2\cdots dw_k \\ &= \frac{n!(k-1)!}{(n+k-1)!}\frac{k(k+1)}{k^2+k+1} \\ &\times \ (1+\frac{(n_1+1)(n_2+1)}{(n+k)(n+k+1)}) \end{aligned} \tag{3.83}$$

It should be noted from (3.81) and (3.83) that we can get back BE statistics with $L \to \infty$ where as for multiple functions $w_1 w_2$ BE statistics is not regained unless one introduces a similar divisor with $w_1 w_2$ which may lead to uniform distribution of $\vec{W}$ in the limiting form. We have considered perturbations in the simplest form. In general one may consider more complex perturbations to obtain similar results. The implications of the " apriori" distribution is explained in more details in a recent paper [Dasgupta 1990]. There we imagine that the transition to indistinguishable elementary particles from distinguishable particles is a continuous process depending on the mass m of the particles ; $L(m) \to \infty$as $m \to 0$. The mass $m \to 0$ means that particles are more and more indistinguishable. The intermediate situation may also arise if one assumes that occupation of a state by a particle may have an influence on the occupation of other states in a special way ; related to it is the screening type effect where the immediate two cells are highly correlated. In the framework of local quantum field theory para Fermi and para Bose statistics essentially reduce to the usual Fermi and Bose statistics and a small violation of the Pauli principle is not admissible if this generalisation is considered to be due to inclusion of the inner degrees of freedom. Although in nature we come across either distinguishable particles following MB statistics or indistinguishable particles following BE/FD statistics , one may always generate mechanism of selection such that the resultant probability distribution is of the type(3.81) i.e. partial indistinguishability.

Let us now look into the correlation structure of w_1 and w_2 for the uniform a priori distribution

$$\operatorname{cor}(w_1, w_2) = -\frac{1}{k-1}$$

To calculate the interference term we need to calculate the correlation function between

$$w_1^{\frac{1}{2}} \text{ and } w_2^{\frac{1}{2}}$$

which is given by

$$\operatorname{cor}(w_1^{\frac{1}{2}}, w_2^{\frac{1}{2}}) = -\frac{\pi}{6k(2k+1)(4-\pi)}[1 + o(1)] \tag{3.84}$$

So the correlation between $w_1^{\frac{1}{2}}$ and $w_2^{\frac{1}{2}}$ approaches zero at a faster rate than between w_1 and w_2. This might have deeper significance in the theory of measurement in quantum mechanics.It is worthmentioning that Kyprianidis et al[Kyprianidis 1987] considered the quantum particles as distinguishable particles within the framework of causal stochastic interpretation of quantum mechanics. Here the particle aspect of the individual micro-objects undergo Feynman like random trajectories in space-time weighted by real positive probability weights. Now the particles are distinguishable by their random trajectories. After averaging over the random weights, the BE statisitcs can be derived taking the same a priori distribution as considered by Kunte and Tersoff and Bayer. The statistical dependence between w_1 and w_2 is thought to be due to non-local correlation in quantum mechanics. This non-local correlations are a manifestation of a fluctuating metric which has been shown to produce the quantum potential[Roy 1986]. After averaging we get the average trajectories of the micro particles and they loose their distinguishable character.

3.7 Quark Confinement :

We have already discussed that a random potential can be generated out of the fluctuation of space-time metric. Dresden [Dresden 1977] tried to construct some random potential in the space-time region of hadronic interaction to confine quarks.
Quantum Chromodynamics (QCD) explains strong interaction where quarks and gluons carrying the "color" quantum number play the essential role in understanding the hadron structure to a certain extent. However, we still lack a good understanding of longer distance hadron physics, in particular when quarks and gluons are "confined" inside hadrons. Many attempts have been made so far in this direction [Efimov 1993].

Namsrai [Namsrai 1993] gave an interesting proposal regarding this confinement. He assumed that the space-time, due to presence of quarks, begins to fluctuate inside the hadrons and its topological structure gives rise to changes in physical parameters of quarks. This means that the structure of space-time is distorted in the immediate neighborhood of a particle (quark), which leads to the concept of confinement and to the problem of reformulating the dynamics of the particle. Here, the random string like objects carry a fluctuation property of space-time and these stochastic strings alter the geometry of space-time at least inside hadrons. This alteration, in turn, affects the behavior of quarks.

Let us consider the stochastic space-time induced by random strings, the behavior of which is described by the probability distribution

$$P[Y] = \frac{1}{N}\exp\{-\frac{1}{2}\int_{M_1}\int_{M_2} d^2\sigma_1 d^2\sigma_2\sqrt{g_1}\sqrt{g_2}Y^{\mu}(\sigma_1)\ D^{-1}_{\mu\nu}(\sigma_1-\sigma_2)Y^{2}(\sigma_2)\} \tag{3.85}$$

where M is a two dimensional surface known as the string world sheet ($\sigma = \sigma^a = \{\sigma^1 = \sigma,\ \sigma^2 = \tau\}$) which is endowed with a metric tensor g_{ab} while space-time has coordinates $x^{\mu(\tau)}$ and metric $G_{\mu\nu}$. $Y^\mu(\sigma)$ are coordinates of strings and N is a normalization constant.

$D^{-1}_{\mu\nu}$ is the inverse of the two-point correlation

$$\langle Y^\mu(\sigma_1)Y^\nu(\sigma_2)\rangle_Y = D^{\mu\nu}(\sigma_1 - \sigma_2) \tag{3.86}$$

For simplicity we take white noise covariance

$$D^{\mu\nu}(\sigma_1 - \sigma_2) = -\eta^{\mu\nu}\frac{\lambda^2}{\sqrt{g}R}\delta^2(\sigma_1 - \sigma_2) \tag{3.87}$$

where R is the Ricci curvature scalar of the manifold M, λ is a constant (as dimension of length). Two possibilities are considered as follows :

(i) $\lambda^2 \sim G$, where G is the Newtonian constant.

(ii) $\lambda^2 \sim \alpha_0'$, α_0' is the inverse string tension.

The condition (i) means that the fluctuation of the string coordinates takes place at the Planck scale while (ii) implies that coordinates $Y^\mu(\sigma)$ obey random properties in a domain characterized by the size of the hadron.

Moreover, the coordinates $Y^\mu(\sigma)$ are assumed to fluctuate around the usual space-time coordinate $x^\mu(\tau)$ and

$$Y^\mu(\sigma) = \frac{1}{\sqrt{g}R}\delta(\sigma^1)x^\mu(\sigma^2 = \tau) + \eta^\mu(\sigma^1, \sigma^2) \tag{3.88}$$

where the functions $\eta^\mu(\sigma)$ are random variables of the type of $Y^\mu(\sigma)$. The transformation

$$\xi^\mu \Rightarrow x^\mu \exp\{\frac{1}{2(\pi\alpha')^{\frac{1}{2}}}\int_M d^2\sigma\sqrt{g}Rn_\mu(\sigma)Y^\mu(\sigma)\} = x^\mu\Lambda \tag{3.89}$$

leads to the metric tensor

$$G_{\mu\nu}(x, Y) = \Lambda^2[\eta_{\mu\nu} + \epsilon_{\mu\nu}(x) + \frac{1}{4}\epsilon^\rho_\mu(x)\epsilon_{\nu\rho}(x)] \tag{3.90}$$

$$G^{\nu\sigma}(x, Y) = \Lambda^{-2}[\eta^{\nu\sigma} - \epsilon^{\nu\sigma}(x) + \frac{3}{4}\epsilon^{\nu\rho}(x)\epsilon^\sigma_\rho(x)] \tag{3.91}$$

where $\epsilon_\mu^\alpha(x) = \frac{1}{\sqrt{\pi\alpha'}} x^\alpha(\tau)\eta_\mu(\tau)$. It can be easily verified that

$$G^{\nu\sigma}(x,Y)G_{\mu\nu}(x,Y) = \delta_\mu^\sigma \tag{3.92}$$

Now considering the averaging over the random variable $Y^\mu(\sigma)$ with the probability distribution as defined earlier, we have

$$\begin{aligned} G_{\mu\nu}(x) &= \langle G_{\mu\nu}(x,Y)\rangle_Y \\ &= [\eta_{\mu\nu} + \epsilon_{\mu\nu}(x) + \frac{1}{4}\epsilon_\mu^\rho(x)\epsilon_{\nu\rho}(x)] \\ &\times \exp\{\frac{\lambda^2}{2\pi\alpha'}\int_M d^2\sigma\sqrt{g}R\} \end{aligned} \tag{3.93}$$

By definition,

$$\text{Euler}(M) = \frac{1}{4}\int_M d^2\sigma\sqrt{g}R = 2 - 2N \tag{3.94}$$

where it is assumed that M has geuus N. So,

$$\begin{aligned} G_{\mu\nu}(x) &= \exp\{\frac{4\lambda^2}{\alpha'}(1-N)\} \\ &\times [\eta_{\mu\nu} + \epsilon_{\mu\nu}(x) + \frac{1}{4}\epsilon_\mu^\rho(x)\epsilon_{\nu\rho}(x)] \end{aligned} \tag{3.95}$$

This implies that the physical space-time metric of the hadronic matter is modified with respect to Minkowski one due to topological properties of space-time at short-distancees. We shall study the constituent quarks in hadrons using the above metric. It is well known that when hadrons travel with velocity $\vec{Q}$, then its proper size α' undergoes Lorentz contraction

$$\alpha' = \alpha^0(1 - \frac{Q^2}{c^2}) \tag{3.96}$$

At the sametime, the constituent quark carries same fraction of the total proton momentum $\kappa v^2 = Q^2$; therefore the topological factor in the above metric $G_{\mu\nu}(x)$ becomes

$$\exp\{\frac{4\lambda^2}{\alpha'}(1-N)\} = \exp\{\frac{4\lambda^2}{\alpha_0^1[1-(\chi\frac{v^2}{c^2})]}(1-N)\} \tag{3.97}$$

where $\vec{v}$ is the velocity of the quark. In the limiting case when the velocity of the quark is small, an additional non-relativistic potential Φ appears

$$\Phi = \frac{1}{2}c^2(-1 - G_{00}) \tag{3.98}$$

Then,

$$\begin{aligned} U &= m\Phi \\ &= -\frac{m}{2}c^2 G_{00} \\ &= \frac{mc^2}{2}\exp\{\frac{4\lambda^2}{\alpha_0'}(1+\frac{\chi\vec{v}^2}{c^2})(1-N)\}[1-\frac{\vec{x}^2}{x_0^2(1-\chi\frac{\vec{v}^2}{c^2})}] \end{aligned} \tag{3.99}$$

where $x_0^2 = 4\pi\alpha_0'$.

Expanding this expression into the power series of $\chi\frac{\vec{v}^2}{c^2}$ and keeping leading terms we obtain

$$U = \frac{mc^2}{2}\{\rho(1-\frac{x^2}{x_0^2}) + \frac{v^2}{c^2}[-2\kappa\rho + L(1-\frac{x^2}{x_0^2})]\} \tag{3.100}$$

with

$$\rho = \exp\{\frac{4\lambda^2}{\alpha_0'}(1-N)\}$$

and

$$L = \chi\rho\{2 + \frac{4\lambda^2}{\alpha_0'}(1-N)\}$$

Namsrai [Namsrai 1993] sketched the form of the quark potential for differential values of N by choosing $\frac{v^2}{c^2} = \frac{1}{9}$, $\chi = 3$ and $\lambda^2 = \alpha_0'$.

The second important charachteristic of the quark is its energy. When the quark moves inside the hadron, its averaged energy is defined as

$$\begin{aligned} E &= \frac{mc^2}{\sqrt{(1-\frac{\vec{v}^2}{c^2})}}\sqrt{-G_{00}} \\ &= \frac{mc^2}{(1-\frac{\vec{v}^2}{c^2})}\exp\{\frac{\lambda^2(1-N)}{\alpha_0[1-\chi\frac{\vec{v}^2}{c^2}]}\} \\ &\times [1-\frac{\vec{x}^2}{x_0^2[1-\chi\frac{\vec{v}^2}{c^2})]}]^{\frac{1}{2}} \end{aligned} \tag{3.101}$$

From this we can conclude that the quark undergoes a finite motion. The phase diagram takes the following form

$$\frac{\vec{p}^2}{p_{\max}^2} + \frac{\vec{x}^2}{x_{\max}^2} \leq 1 \tag{3.102}$$

where $p_{\max} = mc$, $\vec{p} = \sqrt{\chi} m\vec{v}$; $x_{\max} = x_0 = 2\sqrt{\pi\alpha_0'}$. Assuming $\alpha_0' = \frac{1}{m_p^2}$, we get $x_{\max} = 10^{-13} cm$.

In order to illustrate the quark motion from the point of view of the classical theory we assume that a quark-like particles undergoes a finite motion in some domain and that its potential and energy are given by the above formulae.

A quantum mechanical description of the quark motion inside the hadron can be carried out in the standard way. The Hamiltonian function of the non-relativistic motion of the "free" quark can be written as

$$H = (m_q c^2 + \frac{m_q v^2}{2})[1 - \frac{\vec{x}^2}{X_0^2}]^{\frac{1}{2}} \tag{3.103}$$

where the quark mass

$$m_q = m \exp\{\frac{\lambda^2}{X_0[1 - (\chi\frac{v^2}{c^2})]}(1 - N)\}$$

and

$$X_0^2 = x_0^2(1 - \frac{Q^2}{c^2}); \quad x_0^2 = 4\pi\alpha_0'$$

$\vec{Q}$ is the hadron momentum. If the hadron is at rest, $\vec{Q} = 0$. Substituing H and p by quantum mechanical operators, we can get

$$(1 - \frac{\vec{x}^2}{X_0^2})^{-\frac{1}{2}} i\hbar\frac{d\Psi}{dt} = -\frac{\hbar}{2m}\nabla^2\Psi \tag{3.104}$$

and for stationary case

$$\Psi = \text{constant}.e^{(-\frac{i}{\hbar})Et}\Psi(\vec{x})$$

$$\frac{\hbar}{2m}\nabla^2\Psi + (1 - \frac{\vec{x}^2}{x_0^2})^{-\frac{1}{2}} E\Psi = 0 \tag{3.105}$$

It can be easily verified that the function

$$\Psi(x) = \text{constant}\sqrt{1 - \frac{\vec{x}^2}{X_0^2}} \exp\{\frac{i}{\hbar}\vec{p}.\vec{x}(1 - \frac{\vec{x}^2}{X_0^2})^{-\frac{1}{4}}\} \tag{3.106}$$

with $E = \frac{p^2}{2m}$.

So, this solution correctly represents the general physical picture of the given motion of the quark. Indeed the wave function decreases and goes to zero at the surface of the hadron $\vec{x}^2 = x_0{}^2$ and rapidly oscillates.

Now we shall study the general situation when the constituent quark carries some fraction of the total momentum $\chi\vec{v}^2 = \vec{Q}^2$. The total Hamiltonian will be

$$H = D(x^2) + \frac{mv^2}{2}N(x^2) \tag{3.107}$$

It can be written in terms of quantum mechanical operator as

$$\frac{i\hbar(\frac{d\Psi}{dt}) - D(x^2)\Psi}{N(x^2)} = -\frac{\hbar^2}{2m}\Delta\Psi \tag{3.108}$$

where

$$\begin{aligned} D(x^2) &= mc^2[\rho'(1 - \frac{x^2}{x_0^2})^{\frac{1}{2}} + \frac{1}{2}\rho(1 - \frac{x^2}{x_0^2})] \\ N(x^2) &= (1 - \frac{r^2}{r_0^2})^{-\frac{1}{2}}[(\rho' + L^1)(1 - \frac{r^2}{r_0^2}) - 2\chi\rho'] - 2\chi\rho + L(1 - \frac{r^2}{r_0^2}) \end{aligned}$$

Since functions $D(x^2)$ and $N(x^2)$ depend on the radius vector $\vec{x}^2 = r^2$, our problem is reduced to the Schrödinger equation for the quark motion in the central-symmetrical potential. In the stationary case it takes the form

$$\Delta\Psi + \frac{2m}{\hbar^2}[E' - U'(r)]\Psi = 0 \tag{3.109}$$

where $E' = EN^{-1}(r^2)$, $U'(r) = D(r^2)/N(r^2)$.

Using spherical coordinates and writing

$$\Psi = R(r)Y_{lm}(\theta, \phi)$$

we get the following equation for radial functions $R(r)$:

$$\frac{1}{r^2}\frac{d}{dr}(r^2\frac{dR}{dr}) - \frac{l(l+1)}{r^2}R + \frac{2m}{\hbar}[E' - U'(r)]R = 0 \quad (3.110)$$

Outting $R(r) = \frac{\chi(r)}{r}$, the above equation reduces to

$$\frac{d^2\chi}{dr^2} + [\frac{2m}{\hbar^2}(E' - U') - \frac{l(l+1)}{r^2}]\chi = 0 \quad (3.111)$$

The physical requirement of the quark confinement gives

$$\chi(r = r_0) = 0 \quad (3.112)$$

and

$$R \Rightarrow R_{kl} = (-i)^l\sqrt{\frac{2}{\pi}}\frac{r^l}{k^l}(\frac{d}{rdr})^l\frac{\sin kr}{r} \quad (3.113)$$

$k = \frac{p}{\hbar} = \sqrt{\frac{2mE}{\hbar}}$.

at the center of the hadron, $r/r_0 \ll 1$ i.e., it turns to the stationary states of the free quark at small distances at which the quark, at the same as the energy, posseses a definite absolute value of its momentum projection.

Because of the complicated forms of the functions $N(r^2)$ and $D(r^2)$, an explicit solution is very difficult.However, we shall study the energy spectra of excited hadron states numerically. The numerical results can be found to be

$$(E_n^1)_1 = e^{-2}mc^2, \quad (E_n^1)_2 = e^{-2}[mc^2 - \frac{m_p^2c^2}{m}(n + \frac{3}{2})^2] \text{ for } N = 3 \quad (3.114)$$

$$(E_n^2)_1 = \frac{1}{2}e^4mc^2, \quad (E_n^2)_2 = e^4[\frac{1}{2}mc^2 - 36\frac{m_p^2c^2}{m}(n + \frac{3}{2})^2] \text{ for } N = 0 \quad (3.115)$$

$$(E_n^3)_1 = \frac{3}{2}mc^2, \quad (E_n^3)_2 = \frac{3}{2}mc^2 + 25\frac{m_p^2c^2}{m}(n + \frac{3}{2})^2 \text{ for } N = 1 \quad (3.116)$$

We see that fluctuation of the space-time metric around the hadron matter which is caused by the random string field, yields varied spectra of excited hadron states. There exists states with very high energy levels also depending on the Euler number.

Let us now discuss the confinement potential and propagator of the confined particles. The fundamental quantity in the field theory is the propagator or Green function which reflects the causal connection of quantized fields propagating between points of space-time. It is well known that the form of the potential in the static limit is connected with the form of particle propagator. For example :

$$\frac{e}{4\pi r} = \frac{e}{(2\pi)^3}\int d^3pe^{i\vec{p}.\vec{r}}\frac{1}{\vec{p}^2} \tag{3.117}$$

$$\frac{g}{4\pi r}e^{-mr} = \frac{g}{(2\pi)^3}\int d^3pe^{i\vec{p}.\vec{r}}\frac{1}{m^2+\vec{p}^2} \tag{3.118}$$

The confined potential is exactly defined from the total energy of the particle (at rest) as

$$\begin{aligned} U_{\text{tot}} &= D(r^2) \\ &= \{\frac{g_1}{4\pi r_0}(1-\frac{r^2}{r_0^2})^{\frac{1}{2}} + \frac{g_2}{4\pi r_0}(1-\frac{r^2}{r_0^2})\} \quad \text{for } 0 \leq r < r_0 \\ &= 0 \quad \text{for} \quad r_0 < r < \infty \end{aligned} \tag{3.119}$$

where $g_1 = 4\pi r_0 mc^2\rho'$ and $g_2 = 2\pi r_0 mc^2\rho$.

In general, ρ, ρ' and r_0 depend on the external hadron momentum $\vec{Q}$, given by

$$\rho' = \exp\{\frac{\lambda^2}{\alpha_0'(1-Q^2/c^2)}(1-N)\tag{3.120}$$

$$\rho = (\rho')^4, \quad r_0 = [4\pi\alpha_0'(1-Q^2/c^2)]^{\frac{1}{2}} \tag{3.121}$$

We have the integral equation

$$\begin{aligned} \frac{g_1}{4\pi r_0}(1-\frac{r^2}{r_0^2})^{\frac{1}{2}} + \frac{g_2}{4\pi r_0}(1-\frac{r^2}{r_0^2}) &= \frac{c}{2\pi}\int d^3pe^{i\vec{p}.\vec{r}}D_1(\vec{p}^2) \\ &+ \frac{c_2}{(2\pi)^3}\int d^3pe^{i\vec{p}.\vec{r}}D_2(\vec{p}^2) \end{aligned} \tag{3.122}$$

This integral solution has the unique solution

$$\begin{aligned} D(\vec{p}^2) &= g_1\frac{\pi r_0^2}{2}\frac{J_2(r_0\sqrt{\vec{p}^2})}{(r_0\sqrt{\vec{p}^2}} \\ &+ \frac{g_2\sqrt{2\pi}r_0^2 J_{\frac{5}{2}}(r_0\sqrt{\vec{p}^2})}{(r_0\sqrt{\vec{p}^2})^{\frac{5}{2}}} \end{aligned} \tag{3.123}$$

J_2 and $J_{\frac{5}{2}}$ are the Bessel functions of the order 2 and 5/2 respectively. Generalizing this to the relativistic case, we have

$$D(\vec{p}^2) = \frac{\pi}{2} r_0^2 g_1 \left(r_0 \sqrt{-p^2} \right)^{-2} J_2(r_0\sqrt{-p^2}) \tag{3.124}$$

$$+ \sqrt{2\pi} r_0^2 g_2 \left(r_0 \sqrt{-p^2}) \right)^{-\frac{5}{2}} J_{5/2}(r_0\sqrt{-p^2}) \tag{3.125}$$

and similarly $D_M(\vec{p}^2)$ for the particle with mass M (i.e.,gluon like particle with mass M), D_M gives the oscillation potential

$$\begin{aligned} U_M &= \frac{g_1}{2\sqrt{2\pi} r_0^2 \sqrt{M}} (1 - \frac{r^2}{r_0^2})^{\frac{1}{4}} J_{\frac{1}{2}}(M\sqrt{r_0^2 - r^2}) \\ &+ \frac{g_2}{2\pi r_0^2 M} (1 - \frac{r^2}{r_0^2})^{\frac{1}{2}} J_1(M\sqrt{r_0^2 - r^2}) \text{ for } 0 \le r < r_0 \\ &= 0 \text{for } r_0 \ge r > \infty \end{aligned} \tag{3.126}$$

If we assume that force transmitting quanta giving confinement potentials are gluons and Z-bosons-like objects, then

$$\begin{aligned} D(\vec{p}^2) &\Rightarrow G_{\mu\nu} = V_0(-p^2 r_0^2) G^0_{\mu\nu}(p) \\ D_M(\vec{p}^2) &\Rightarrow G^M_{\mu\nu}(p) = V_M(-p^2 r_0^2) G^{0M}_{\mu\nu}(p) \end{aligned} \tag{3.127}$$

where

$$\begin{aligned} V_M(-p^2 r_0^2) &= \frac{\pi}{2} J_2(r_0\sqrt{M^2 - p^2} \\ &+ \frac{\pi}{2p'^3} (r_0\sqrt{M^2 - p^2})^{-\frac{1}{2}} J_{\frac{5}{2}}(r_0\sqrt{M^2 - p^2}) \end{aligned} \tag{3.128}$$

$$V_0(-p^2 r_0^2) = V_M(-p^2 r_0^2) \mid_{M=0} \tag{3.129}$$

$G^0_{\mu\nu}(p^2)$ and $G^0_{\mu\nu}{}^M(p^2)$ are the local propagators of gluons and massive gluons respectively. We call the latter colour Z-like bosons. The propagators

$$G^0_{\mu\nu}(\vec{p}^2) \text{ and } G^0_{\mu\nu}{}^M(\vec{p}^2)$$

posses the interesting properties as

- (i) They are entire analytic functions of the variable p^2.
- (ii) Their order of growth is $\rho = 1/2$.
- (iii) They decrease rapidly enough in the Euclidean directions $p^2 \Rightarrow -\infty$.
- (iv) They have no poles in the complex plane p^2, so that force-transmitting quanta corresponding to these propagators are never observable.
- (v) Wave functions of these quanta are spread out over space-time,

$$G_{\mu}^{\nu\,\text{nonlocal}}(x) = \int d^4y K_0(x-y) G_{\mu}(y) \qquad (3.130)$$

or

$$Z_{\mu}^{\text{nonlocal}}(x) = \int d^4y K_m(x-y) \hat{Z}_{\mu}(y) \qquad (3.131)$$

where $K_0(x)$ and $K_m(x)$ are nonlocal generalized functions, the Fourior tranforms of which are defined by $| V_0(r_0^2 p^2) |^{\frac{1}{2}}$ and $| V_m(r_0^2 p^2) |^{\frac{1}{2}}$, respectively. The local wave function of gluons $G_{\mu}(x)$ and color Z-like bosons $\hat{Z}_{\mu}(x)$ yield the usual propagators

$$\begin{aligned} \langle | T\{G_{\mu}(x) G_{\nu}(y)\} | 0\rangle &= \frac{1}{i}\Delta^{0}_{\mu\nu}(x-y) \\ \langle | T\{\hat{Z}_{\mu}(x) \hat{Z}_{\nu}(y)\} | 0\rangle &= \frac{1}{i}\Delta^{M}_{\mu\nu}(x-y) \end{aligned} \qquad (3.132)$$

by means of their T-product operators in x-space. Concerning the quark propagator, it should be constructed by the same procedure as has just been defined for gluon one. The confinement propagator for the quark can be written as

$$\hat{S}_R(p) = V_m(-p^2 r_0^2) \hat{S}^0(p) \qquad (3.133)$$

where $\hat{S}^0(p)$ is the usual local propagator of the quark, which is used in QCD. Thus we see that the confinement potential due to the induced gravitation of "hadron" strings leads to a new type of strong interaction in which there exists two decreasing coupling constants of $g_1 = 4\pi r_0 m_L c^2$ and $g_2 = 2\pi r_0 m_q^4 c^2/m^3$ at $N \Rightarrow \infty$ and $| Q | \Rightarrow c$. The corresponding force transmitting quanta become nonlocal and unobservable, and their propagators are entire analytic functions.

NON-LOCAL FIELDS

"What is usually called rigid body in non relativistic quantum mechanics is nothing but an idealized case of many particle system ... if we go over to relativistic quantum theory, we have to deal also with systems which have finite degrees of freedom .... This gives rise to the year old problem of divergence which in turn, brings about the following question. In what way can the atomic structure of matter and energy be reconciled with the assumption of unlimited divisibility of space and time"

H. Yukawa : Atomistics & the Divisibility of Space & Time.

4. NON-LOCAL FIELDS

In chapter 1, we have already mentioned that Schweizer and Sklar [Schweizer 1983] constructed distribution generated spaces and hence C-spaces during their investigations to overcome the difficulties raised by Wald [Wald 1952] within Menger's framework. During that period Rosen [Rosen 1947] independently initiated a discussion on the model of elementary particles and statistical geometries. In both the approaches, hazy lumps are considered instead of point set and hazy lump may represent the seat of elementary particles. So, instead of point particles, concept of extended fundamental objects in space-time are introduced in connection with statistical geometry. In a sense, a kind of non-locality is introduced to understand the behavior of microparticles and the space-time structure at small distances.

So far various authors considered two different approaches to the construction of non-local theory. In one approach, it is assumed that there exists a fundamental constant of the dimension of length together with the constant like the velocity of light c and Planck's constant h. One may envisage dramatic changes of our concepts of physical world like concept of space-time, concept of causality etc. within this fundamental length. A comprehensive review regarding the introduction of fundamental length can be found in the monograph published by Namsrai [Namsrai 1986]. On the otherhand, in the second approach this length scale is not a constant one. It characterizes the domain of non-local interaction or the size of the extended particles, say, of electron, proton etc. Here, the boundaries of the extended objects may not be sharply defined. Efimov [Efimov 1993] elaborated this idea in a series of recent publications. The recent experiments [Bracci 1983] have been designed to probe the distances up to 10^{-17} cm and leptons are also considered as of extended structure. In this chapter we shall concentrate our discussion on the non-locality of fields considering the extended structure of elementary particles but without any fixed boundary so as to be in conformity with the above idea of statistical geometry.

4.1. Extended Elementary Objects :

In early fifties Yukawa [Yukawa 1950] developed the idea of non-local fields considering the extended structure of elementary particles. The principal idea behind his approach is that one need two variables to define an extended object. If we consider two points in space-time, say, x_1 and x_2 in one dimension, then we can define $x = \frac{x_1+x_2}{2}$ and $\xi = x_1 - x_2$ to characterize this extended object. The variable x can be used to define the mean motion of the object and ξ is known as internal variable. In that case, the field functions will be functions of both the variables x abd ξ. This is usually known as non-local fields. But non-locality may also arise from the non-local interaction of the local fields. Subsequent to the publication of Yukawa's papers lot of works have been done in case of both kind of non-locality. In sixties, lot of enthusiasm arose to establish a connection of this internal variable with quantum numbers like Isospin, Strangeness, Baryon number etc. within this framework of non-local fields. de Broglie [de Broglie 1963] published a monograph called "....Vigier theory of elementary particles" discussing most of these results. Several attempts have been made to construct a unified theory of elementary particles based on the assumption that each particle has an extended structure in the space-time world.However, such kind of attempts face the problem of interrelation as follows .

One of the main points in general theory of relativity is to establish the interrelation between the geometry of large scale physical world and the distribution and motion of matter in it. If we assume that such kind of interrelations also exist in the microdomain, those which are obtained on formal change of classical Einstein equations by quantizing gravitational field together with matter field must be relatively unimportant, because the characteristic length which is constructed from the gravitational constant,

Planck constant and the velocity of light is very small compared with the length characterizing the scale of the world of elementary particles. Therefore, if we intend to resurrect the spirit of general relativity in the world of elementary particles, we have to look for the possibility of finding interrelations between the geometrical properties, in a general and abstract sense, of the world of elementary particles and their more directly observable properties. At present there is no conclusive proof that gravitational field can be quantized in a consistent way. So we can take gravitational field as kind of classical field. Now contrary to the case of general relativity dealing with classical field, the unlimited divisibility of space and time into smaller parts may well become the question in the theory of elementary particles. During the advancement of quantum mechanics, most physicists started to think that the contradiction of the atomic structure of matter and energy and the continuity of space-time disappeared forever by unifying the particle and wave aspects of matter and energy. In nonrelativistic quantum mechanics, it may be a good proposition. But in case of relativistic quantum mechanics, we deal with system which have infinite degrees of freedom. This gives rise to the year-old problem of divergence, which, in turn, brings about the following new question :

"Inwhat way can the atomic structure of matter and energy be reconciled with the assumption of unlimited divisibility of space-time on subnuclear level?"

In nonrelativistic quantum mechanics the compatibility of the discreteness of matter and energy with the continuity of space-time is usually achieved by introducing the second quantization formalism. Let $\Psi(x,t)$ and $\Psi^+(x,t)$ be the scalar field operator and its Hermitian conjugate. Then in the second quantization formalism, they satisfy the commutation relations

$$\begin{aligned} [\Psi(x,t), \Psi^+(x',t)]_\pm &= \delta(x-x') \\ [\Psi(x,t), \Psi(x',t)]_\pm &= 0 \\ [\Psi^+(x,t), \Psi^+(x',t)]_\pm &= 0 \end{aligned} \tag{4.1}$$

where x stands for three space coordinates and $\pm$ discreminates between bosons and fermions. $\Psi(x,t)$ is assumed as annihilation operator and $\Psi^+(x,t)$ as creation operator for such a particle. The formal extension can be done in the relativistic domain assuming the local character of commutation relations for the local field. This approach presupposes the decomposition of matter and energy into particles and quanta which can be localized at a point at ay instant while the continuity of space-time is taken for granted. At this point Yukawa [Yukawa,1966] proposed a novel approach which is in compatible with his basic idea of non-local field. We can construct the operator on small three dimensional space domain D as

$$\begin{aligned}\Phi(D,t) &= (\frac{1}{\sqrt{\sqrt{D}}}) \int_D \Psi(x,t)dx \\ \Phi^+(D,t) &= (\frac{1}{\sqrt{\sqrt{D}}} \int_D \Psi^+(x,t)dx \end{aligned} \tag{4.2}$$

which satisfy the following commutation realtion

$$[\Phi(D,t), \Phi^+(D,t)]_\pm = 1$$

where $\int_D$ denotes the integration over the domain with the value V_D. Now if we consider another domain D' and define the operators $\Phi(D',t)$ and $\Phi^+(D',t)$ we have the commutation relations

$$\begin{aligned}[\Phi(D,t), \Phi^+(D',t)]_\pm &= \frac{V_{DD'}}{\sqrt{V_D V_{D'}}} \\ [\Phi(D,t), \Phi(D,t)]_\pm &= 0 \\ [\Phi^+(D,t), \Phi^+(D',t)]_\pm &= 0\end{aligned} \tag{4.3}$$

where $V_{DD'}$ is the volume of the domain $D \cap D'$ common to D and D'.

If the domain D does not overlap with D', we have

$$[\Phi(D,t), \Phi^+(D',t)]_\pm = 0 \tag{4.4}$$

In the spirit of defining $\psi(x,t)$ and $\Psi(x',t)$, we can interpret $\Phi(D,t)$ and $\Phi^+(D,t)$ as annihilation and creation operators. The number operator can be defined as

$$n(D,t) = \Phi^+(D,t)\Phi(D,t) \tag{4.5}$$

which commutes with $n(D',t)$ defined for any domain D' other than D. It should be observed that these number operators and their commutation relations are different from our previous one (in terms of local field operators). But it is necessary to impose some restrictions on the size and shape of the domains. Simply we may consider the spherical domains of radius to which can be located arbitrarily in space. These are known as elementary domain[Yukawa,1966]. In this picture, an elementary domain may be considered as the seat of an elementary particle. Now if we consider a system consisting of any number of elementary particles in this sense, which are contained in a very large box, there are only finitely many number of operators $n(D_1,t), n(D_2,t)\cdots n(D_N,t)$ which commute with each other. Here, the elementary domains $D_1, D_2, \cdots, D_N$ are assumed as non-overlapping domains. If we consider further $n(D',t)$, of an arbitrary elementary domain which does not belong to the original set $\{D_1, D_2, \cdots, D_N\}$, $n(D',t)$ can not be commutative with all of $n(D_1,t), (D_2,t), \cdots, n(D_N,t)$. Otherwise $n(D',t)$ has to be included in the set $\{n(D_1,t), (D_2,t), \cdots, N(D_N,t)\}$. In this sense, the set $\{D_1, D_2, \cdots, D_N\}$ can be called a complete set of elementary domains. Such a complete set is obtained by filling the whole space in the box with elementary spheres which do not overlap with each other, any further addition of elementary spheres being impossible without overlapping with those already in the box. There exists many possibilities in choosing a complete set. If we define the center of any domain D by three coordinates

$$X = \frac{1}{\sqrt{V_D}} \int_D x dx$$

the complete set $\{D_1, D_2, \cdots, D_N\}$ is charaterized by $3N$ parameters $X^{(1)}, X^{(2)}, \cdots, X^{(N)}$. The following conditions are assumed,

$$\left|X^{(r)} - X^{(s)}\right| \geq 2l^0; \qquad r, s = 1, 2, \cdots, N \tag{4.6}$$

where

$$X^{(r)} = \frac{1}{\sqrt{V_0}} \int_{D_r} x dx \quad \text{with} \quad V_0 = 4\pi \frac{l_0^3}{3}$$

Here, $\mid X^{(r)} - X^{(s)} \mid$ denotes the distance between two points $X^{(r)}$ and $X^{(s)}$.

Now, we may assume

$$\begin{aligned} \Phi'(D'_\rho, t) &= \textstyle\sum_r C_{\rho r} \Phi(D_r, t) \\ \Phi^{+'}(D'_\rho) &= \textstyle\sum_r C^*_{\rho r} \Phi^+(D_r, t) \end{aligned} \tag{4.7}$$

where $C^*_{\rho r}$ is the complex conjugate of $C_{\rho r}$. It is possible to derive also the commutation relations

$$\begin{aligned} [\Phi(D_r, t), \Phi^+(D_s, t)]_\pm &= \delta_{rs} \\ [\Phi(D_r, t), \Phi(D_s, t)]_\pm &= 0 \\ [\Phi^+(D_r, t), \Phi^+(D_s, t)]_\pm &= 0 \end{aligned} \tag{4.8}$$

as well as for $\Phi'(D'_\rho, t)$ and $\Phi'^+(D'_\rho, t)$. Then we get

$$\sum_r C_{\rho r} C^*_{\sigma r} = \delta_{\rho r}, \quad \rho,\ \sigma = 1, 2, \cdots, N \tag{4.9}$$

In this approach, the free particle equations (for point particles) should be modified in case of these type of extended particles. The Hamiltonian operator H is assumed to have the form

$$H = \sum_{r,s} \phi_r^+ H_{rs} \Phi_s \tag{4.10}$$

where $\Phi_r = \Phi(D_r, t)$, $\Phi_r^+ = \Phi^+(D_r, t)$ and H_{rs} constitute a Hermitian matrix with N-rows and columns and Δt is a very small time interval of the order of l_0/c. In the limit $l_0 \to 0$, we can get the usual Hamiltonian by choosing the matrix element H_{rs} such that it does not vanish. more precisely, for two spheres D_r, D_s with

$$2l_0 \leq | X^{(r)} - X^{(s)} | < 4l_0 \tag{4.11}$$

one can take $H_{rs} = -H_{sr}$, which corresponds to a product of $\Phi^+(x, t)$ and the first derivative of $\Phi(x, t)$ in the direction of the vector $(X^{(r)} - X^{(s)})$ in the limit $l_0 \to 0$.

This model has great similarity with classical model in the dynamical theory of gases. In dynamical theory of gases, we take large number of small rigid spheres with same radius in a box. They are supposed to collide with each other and with the wall of the box according to classical laws of dynamics. In the above model of elementary domains, the fermions (which corresponds to anticommutativity) resemble the classical model more than boson case because the impenetrability of matter is retained in the former case. However, there are also important differences between Yukawa model and the classical model. In the classical gas model, the space in the box which is not occupied by the rigid sphere is simply an empty space whereas the whole space in the box is covered by elementary spherical domains irrespective of whether they are occupied or empty. This is due to the fundamental difference between classical and quantum mechanics. The existence and nonexistence of an object at a certain position, at a certain instant of time can be determined unambigously in classical dynamics. On the otherhand, they are considered simultaneously with respective probability, in general , in quantum mechanics. Thus, even when a domain becomes empty, we may imagine that the domain itself remains from the quantum theoretical point of view. Inspite of such differences, the similarity between these two models help us in modofying the model of elementary domains.

This model can be modified in several ways. One important modification has been done by considering the oriented elementary domains. Here, each elementray domain is supposed to be characterized by three Euler angles $\theta_1, \theta_2, \theta_3$ defining its orientation in addition to three coordinates X_1, X_2, X_3 defining its center. Further, it is assumed that the operators $\Phi(X_1, X_2, X_3; \theta_1, \theta_2, \theta_3), \Phi^+(X_1, X_2, X_3; \theta_1, \theta_2, \theta_3)$ ascribed to each domain where the centers of two domains coincide, but the orientations are different.

Now a system can be described by the set of the operators

$$\Phi\left(X_1^{(r)}, X_2^{(r)}, X_3^{(r)}; \theta_1(r), \theta_2^{(r)}, \theta_3^{(r)}, t\right)$$

and

$$\Phi^+\left(X_1^{(r)}, X_2^{(r)}, X_3^{(r)}; \theta_1^{(r)}, \theta_2^{(r)}, \theta_3^{(r)}, t\right)$$

with $r = 1, 2, 3, \cdots, N$ and $X^{(1)}, X^{(2)}, \cdots, X^{(N)}$ as mutually restricted by the relations

$$| X^{(r)} - X^{(s)} | \geq 2l_0; \quad \text{for} \quad \neq s$$

and each set of $\{\theta_1^{(r)}, \theta_2^{(2)}, \theta_3^{(3)}\}$ are continuous parameters identical with the Euler angles

$$0 \leq \theta \leq \pi;\ 0 \leq \Psi < 2\pi; 0 \leq \Phi < 2\pi$$

It is well known that the wave function of a rigid body in quantum mechanics can be expanded into the series of eigenfunctions

$$U_{JKM}(\theta_1, \theta_2, \theta_3)$$

where J either a positive integer including zero or a positive half integer and

$$K = J, J-1, \cdots, -J-1, -J$$

$$M = J, J-1, \cdots, -J-1, -J$$

The absolute square of angular momentum is $\hbar^2 J(J+1)$ and$K$ and M are its third components in the body frame and the laboratory frame respectively. The eigen function $U_{JKM}(\theta_1, \theta_2.\theta_3)$ is either one valued or two valued as J is integer or a half integer. We can expand Φ and Φ^+ in terms of spherical basis functions. Now if U_{JKM} is properly normalized, we obtain the commutation relations

$$\begin{aligned} [\Phi_{JKM}(\tilde{X}^{(r)},t),\Phi^+_{J'K'M'}(\tilde{X}^{(s)},t)]_\pm &= \delta_{rs}\delta_{JJ'}\delta_{KK'}\delta_{MM'} \\ [\Phi_{JKM}(\tilde{X}^{(r)},t),\Phi_{J'k'M'}(\tilde{X}^{(s)},t)]_\pm &= 0 \end{aligned} \tag{4.12}$$

Here, U_{JKM} and U^+_{JKM} can be interpreted as the annihilation and creation operators of a particle which occupies the domain $D^{(r)}$ and is rotating with the spin angular momrntum designated by J, K, M.

Another modification can be done in the following manner. The operators $\Phi(D)$ and $\Phi^+(D)$ can be generated to

$$\begin{aligned} \Phi(f(x),t) &= \int_D f(x)\Psi(x,t)dx \\ \Phi^+(f(x),t) &= \int_D f^+(x)\Psi^+(x,t)dx \end{aligned} \tag{4.13}$$

where $f(x)$ is an arbitrary function defined in the domain D. If we introduce the spherical coordinates (r,θ,ϕ) with its origin at the center X of the domain D, $f(x)$ can be expanded into series of spherical harmonics

$$f(x) = \sum_{l,m} f_{lm}(r)Y_{lm}(\theta,\phi)$$

and the new set of operators

$$\Phi_{lm}(X^{(r)},t) = \frac{1}{g_{lm}}\int_{D_r} f_{lm}(r)Y_{lm}(\theta\phi)\Psi(x,y)r^2\sin\theta drd\theta d\phi \tag{4.14}$$

and their Hermitian conjugates for elementary spherical domains $D_1, D_2, \cdots, D_N$ satisfy the commutation relations

$$\begin{aligned} [\Phi_{lm}(X^{(r)},t),\Phi^+_{l'm'}(X^{(r)},t)]_\pm &= \delta_{rs}\delta_{ll'}\delta_{mm'} \\ [\Phi_{lm}(X^{(r)},t),\Phi_{l'm'}(X^{(r)},t)]_\pm &= 0 \end{aligned} \tag{4.15}$$

where

$$g_{lm} = \sqrt{\int_0^{l_0} f^+_{lm}(r)f_{lm}(r)r^2dr} \tag{4.16}$$

Thus $\Phi_{lm}(X^{(r)},t), \Phi^+_{lm}(X^{(r)},t)$ can be interpreted as the annihilation and creation operators for a particle with the radius l_0, the center being located at $X^{(r)}$ and the spin angular momentum specified by l, m. One advantage of this model is that in addition to rotational motion one can get vibrational motion. But the problem is that the spin of the particle is restricted to zero or an integer. This model can be further generalized so as to generate the different quantum numbers chracterizing the elementary particles [de Broglie, 1962].

In local field theories, each point in space is the seat which a point particle can occupy or else is empty. Thus there is no need to discriminate between three dimensional space continuum and the whole set of seats for the particles. Now if we consider that each of the elementary particles is an entity with spatially extended structure, each of the seats for it is to have also a finite volume. We have already pointed out that there is no clean-cut distinction between the empty and occupied seats, but they are probabilistically connected with each other, if the entity is a quantum theoretical object. In such a case, the totality of seats remains irrespective of which seats are empty or occupied and is to be discriminated from the space continuum. The discreteness of matter and energy and the unlimited divisibility of space can be reconciled with each other in this way. It should be emphasized that Menger tried to resolve Poincaré dilemma by considering the set of hazy lumps and introducing the notion of classical probability (C-space) at the microlevel. Here, in Yukawa,s picture we get similar idea but using the quantum theretical concepts. It requires serious efforts to understand this type of structure so as to get unified theory of elementary particles with consistent theory of non-local fields.

In recent time, new interest has arisen in connection with the unified description of all interactions where strings and superstrings are considered as extended objects [Green 1986]. This renews the interest to consider the role of fundamental length in understanding the microlocal structure of space-time. Namsrai [Namsrai 1996] made an attempt to define this fundamental length by using the physical characteristics of the electric charge. It is worth mentioning that in early seventies, the present author along with his collaborators [Bandopadhyaya 1976] tried to understand this kind of electric charge distribution over space-time within the framework of non-local fields. Here, each elementary domain of space-time is considered as seat of a massive electron like electron or muon. This lepton derives mass and charge from the system of non-local interaction among massless particles inside the domains. So the system of non-local interaction comprising the massive particle moves with the velocity smaller than that of its massless constituents (which always move with the speed of light). Here, the space-time within the quantized domain is taken as anisotropic to have a consistent theory of non-local field. This anisotropy effect can be expected also by the introduction of directional dependence into the space-time as a kind of Finslerian structure. In this picture, we have introduced a form factor characterizing the non-local interaction and ultraviolet divergence has been removed by using the fundamental length as cut-off. On the otherhand, Namsrai used the idea of a smeared delta function involved in the definition of the idealized concept of a point like charge. Here,

$$e\delta(\vec{r}) \to e\rho_l(\vec{r}) \quad \text{and} \quad \lim_{\ell\to 0} \rho_\ell(\vec{r}) = \delta(\vec{r})$$

e being the elementary point charge and $\rho_\ell(\vec{r})$ describing the extended electric charge due to the existence of the fundamental length. It is assumed that the charge distribution $\rho_\ell(\vec{r})$ has universal characteristics and is independent of the concrete properties of elementary constituents (say, electron, quark etc.) which carry the electric charge. This leads to "non-local" Poisson equation

$$\Delta\phi_\ell(\vec{r}) = -e\rho_\ell(\vec{r}) \tag{4.17}$$

The solution can be written as

$$\phi_\ell(\vec{r}) = \frac{e}{4\pi}\int d\vec{r}'\frac{\rho_\ell(\vec{r}-\vec{r}')}{|\vec{r}|} \tag{4.18}$$

This is a modified form of the Coulomb law at short distances. Modification of Coulomb law gives rise to a change of electromagnetic field as

$$A_\mu(x) \Rightarrow A_\mu{}^\ell(x) \Rightarrow \rho_\ell(\vec{r})A_\mu(x) \tag{4.19}$$

This formalism allows us to introduce the smeared out electromagnetic field $A_\mu{}^\ell(x)$ as

$$A_\mu{}^\ell(x) = \rho_l^2 \otimes A_\mu(x) = \int d^4y\rho_\ell^2(x-y)A_\mu(y) \tag{4.20}$$

Generally speaking, $\rho_\ell(x)$ is a generalized charge density in Minkowski space. In Euclidean metric $\rho_\ell(x_E)$ possesses a probability measure satisfying the condition

$$\int d^4x_E\rho_\ell^2(x_E) = 1 \tag{4.21}$$

and the photon propagator will be

$$D_{\mu\nu}(x-y) = \frac{1}{(2\pi)^4}g_{\mu\nu}\int d^4p\exp[ip(x-y)]\frac{[\tilde{\rho}_\ell(p)]^2}{-p^2-i\epsilon} \tag{4.22}$$

where, $p^2 = p_0^2 - |\vec{p}|^2$, and $\bar{\rho}^2(p)$ is the Fourier transform of the generalized density $\rho_\ell^2(x)$. Thus the expected charge distribution $\rho_\ell(\vec{r})$ should obey the following condition

$$\bar{\rho}_\ell(\vec{p}) = [\bar{\rho}_\ell^2(\vec{p}, p_0)]^2 \,|_{p_0 \Rightarrow 0} \tag{4.23}$$

in the static limit.

Theorem : The non-local charge distribution of the Gaussian form

$$\rho_\ell(\vec{r}) = \frac{1}{\pi^{3/2} l^3} \exp\left(-\frac{\vec{r}^2}{l^2}\right) \tag{4.24}$$

and its Euclidean extension

$$\rho_\ell^2(x_E) = \frac{4}{\pi^2 l^4} \exp\left[-\frac{2}{l^2}(|\vec{x}|^2 + x_4^2)\right] \tag{4.25}$$

satisfy all the above conditions. This is a modified form of the Coulomb law at short distances.

A non-local gauge transformation connected with the extended charge allows one to construct gauge invariant non-local quantum electrodynamics, free of ultraviolet divergencies.

Electromagnetic processes of the type

$$e^- e^- \to e^- e^-$$

$$e^+ e^- \to e^+ e^-$$

$$e^+ e^- \to \mu^+ \mu^-$$

have been studied in this framework. The ratio of cross-sections calculated by local and non-local theories is found to be

$$\frac{\sigma_{\text{nonlocal}}}{\sigma_{\text{local}}} \sim 1 + 2\nu(1) s l^2 \tag{4.26}$$

where $s = (p_1 + p_2)^2 = (2E)^2 = W^2$ and $W = 2E$ as the total energy in the center of mass system. Taking the experimental data [Bartel 1980] , [Berger 1980), it gives

$$l \le 10^{-16}\text{cm}$$

However, local quantum electrodynamics is the best tested theory. So far no deviation has been found even at small distances. Tests of locality are usually performed by using very high precision experiments in physics, considering the experimental values of the anomalous magnetic moment of the electron and muon [Berger 1980],

$$\begin{aligned} \Delta\mu^{e}_{\mathrm{exp}} &= 1.001159652193 \pm 0.000000000010 \\ \Delta\mu^{\mu}_{\mathrm{exp}} &= 1.001165923 \pm 0.000000008 \end{aligned} \tag{4.27}$$

which is in full agreement with QED. Taking these corrections, we get the bounds on fundamental lengths as

$$\begin{aligned} l_e &\leq 8.8 \times 10^{-15}\mathrm{cm} \ \text{ for } \ \Delta^{e}_{\mathrm{exp}} \\ l_n &\leq 1.2 \times 10^{-15}\mathrm{cm} \ \text{ for } \ \Delta^{\mu}_{\mathrm{exp}} \end{aligned} \tag{4.28}$$

So, these are the radii of the electron & muon respectively.

4.2 Internal variable and fluctuation of space-time metric :

Now, let us discuss the fluctuations of space-time in connection with the internal variable or degrees of freedom of elementary particles. The fluctuation of space-time is represented by the probability distribution function and its probability amplitude characterizes the internal freedom of elementary particles. So, the variety of elementary particles reflects the variety of contributions of space-time to natural phenomena.

In the high energy scattering, we may select the probability amplitude in such a way that it may describe the variety of elementary particles and at the same time, that the resulting propagation function may be well regularized. This idea is different from the conventional image i.e., the probability distribution of matter in a space-time of unlimited divisible structure. Our concept of space-time presupposes the possibility of new geometry with fluctuation. The extended structure of elementary particles is never compatible with the unlimited divisible structure of space-time, but with the limited divisible structure of space-time, as is understood from the consideration of observation.

In chapter 3, we have considered fluctuation of the fifth dimension and it gives rise to the quantum potential. Our fluctuating four-dimensional space is structurally the five-dimensional probability distribution of four-dimensional subspaces corresponding to various values of a as in equation (3.44). The metric fundamental form τ' can be defined as [Takano 1967]

$$\tau'^2 = \tau^2 - (\zeta_2 - \zeta_1)^2 \tag{4.29}$$

where

$$\tau^2 = c^2(t_2 - t_1)^2 - (x_2 - x_1)^2 - (y_2 - y_1)^2 - (z_2 - z_1)^2$$

is the usual fundamental form in 4-dimensional space-time. In quantum field theory, the $\overline{\Delta}$-function can be given by the weighted average of Δ in the relevant subspaces :

$$\overline{\Delta}'(\tau, k_m) = \int\int_{-\infty}^{\infty} \pi(\zeta_2)\overline{\Delta}(\tau', k_m)\pi(\zeta_1)d\zeta_1 d\zeta_2 \tag{4.30}$$

By means of the relation

$$\begin{aligned}\overline{\Delta}(\tau', k_m) &= \int \overline{\Delta}(\tau, k)[\delta(k^2 - k_m^2) - \frac{1}{2}(\zeta_2 - \zeta_1)^2 \\ &\times \frac{Re\ H_1(\sqrt{(\zeta_2 - \zeta_1)^2}\sqrt{k^2 - k_m^2})}{\sqrt{(\zeta_2 - \zeta_1)^2}\sqrt{k^2 - k_m^2}}]dk^2\end{aligned} \tag{4.31}$$

$\overline{\Delta}$-function is given as follows :

$$\overline{\Delta}'(\tau, k_m) = \int \overline{\Delta}(\tau, k)\rho(k^2)dk^2 \tag{4.32}$$

where

$$\rho(k^2) = \int_0^{\infty} [\delta(k^2 - k_m^2) - \delta(k^2 - k_m^2 - \lambda^2)]G(\lambda)d\lambda \tag{4.33}$$

Here,

$$G(\lambda) = \int\int_{-\infty}^{\infty} \pi(\zeta_2)\sqrt{(\zeta_2 - \zeta_1)^2} \times J_1(\sqrt{((\zeta_2 - \zeta_1)^2\lambda})\pi(\zeta_1)d\zeta_1 d\zeta_2 \tag{4.34}$$

with $\int G(\lambda)d\lambda = 1$.

The probability distribution $\pi(\zeta)$ is equal to the absolute square of the probability amplitude $\phi(\zeta)$ characterizing the internal freedom of elementary particles. Similar formulae for $\Delta^{(i)}$ and Δ-functions may also be derived :

$$\Delta^{(1)\prime}(\tau, k_m) = \int \Delta^{(1)}(\tau, k)\rho(k^2)dk^2 \tag{4.35}$$

$$\Delta'(\tau, k_m) = \int \Delta(\tau, k)\rho(k^2)dk^2 \tag{4.36}$$

Now it is evident from the form of spectral function $\rho(k^2)$,that the indefinite metric comes out and the divergence difficulty can be removed.

The above formulation can be generalized to the case of multidimensional fluctuation. The metric fundamental form τ' in each distributed subspace is related to the usual form τ in the following way :

$$\tau'^2 = \tau^2 - (\rho_1^2 + \rho_2^2 - 2\rho_1\rho_2 \cos\theta) \tag{4.37}$$

where ρ_1 and ρ_2 are the radii and θ is the angle between them. Here, $(\rho_1^2+\rho_2^2-2\rho_1\rho_2\cos\theta)$ plays the part of $(\zeta_2-\zeta_1)^2$ in the previous fundamental form τ'. The factor

$$\sqrt{\rho_1^2 + \rho_2^2 - 2\rho_1\rho_2\cos\theta}\, J_1(\lambda\sqrt{\rho_1^2 + \rho_2^2 - 2\rho_1\rho_2\cos\theta} \tag{4.38}$$

will come out in the integral expression of $G(\lambda)$.

In case of three-dimensional harmonic oscillator, we can follow a simpler procedure. For this, let us take the Hermite polynomial of order 0 as the probability amplitudes i.e.,:

$$\phi(\zeta,\eta,\xi) = (k_0^{1/2}/\pi^{1/4})^2 \exp[-k_0^2(\zeta^2+\eta^2+\xi^2)/2] \tag{4.39}$$

Then, $G(\lambda)$ will be expressed as :

$$\begin{aligned} G(\lambda) \;=\; & (\frac{k}{\sqrt{\pi}})^2 \int\int\int\int\int\int_{-\infty}^{\infty} \sqrt{(\zeta_2-\zeta_1)^2 + (\eta_2-\eta_1^2) + (\xi_2-\xi_1)^2} \\ & \times J_1(\lambda\sqrt{(\zeta_2-\zeta_1)^2 + (\eta_2-\eta_1)^2 + (\xi_2-\xi_1)^2}) \\ & \times \exp\{-k_0^2[(\zeta_2^2+\zeta_1^2) + (\eta_2^2+\eta_1^2) + (\xi_2^2+\xi_1^2)]\} \\ & d\zeta_1 d\zeta_2 d\eta_1 d\eta_2 d\xi_1 d\xi_2 \end{aligned} \tag{4.40}$$

By suitable transformations we obtain

$$G(\lambda) = \left(\frac{3\lambda}{2k_0^2}\right), \quad F_1(5/2, 2 : -\lambda^2/2k_0^2) \tag{4.41}$$

where, F_1 is confluent hypergeometric functions.

The spectral function $\rho(k^2)$ associated to the above $G(\lambda)$ is shown to satisfy the conditions of regularization given by Pauli and Villars :

$$\int \rho(k^2)dk^2 = \int \rho(k^2)k^2 dk^2 = 0 \tag{4.42}$$

Within this framework, the modified Coulomb potential can be written as

$$V_c' = \sqrt{(\frac{2k_0}{\pi})(\frac{k_0^2 r^2}{4})\exp(\frac{k_0^2 r^2}{4})} \times [k_1(\frac{k_0^2 r^2}{4}) - k_0(\frac{k_0^2 r^2}{4})] \qquad (4.43)$$

which is not singular and equal to $\sqrt{\frac{2k_0}{\pi}}$ at the origin and behaves likes $\frac{1}{r}$ at infinity.

4.3 Quantum Geometry :

We have already discussed that the physical space is described by "statistical geometry" in which the elements are not points but clouds and hence the name C-space. Rosen [Rosen, 1962] tried to elaborate this idea for the description of elementary particles. He called this as " quantum geometry". In this geometry a correspondence can be established between the Cartesian coordinates of a point $\vec{\xi} = (\xi, \eta, \zeta)$ in an abstract space and a probability density for the distribution of the Cartesian coordinates $\vec{X} = (x, y, z)$ in ordinary space. If the space is assumed to be homogeneous and isotropic and the results of the measurements of $x, y,$ and z are dependent of one another one arrives at a Gaussian distribution function

$$\chi(\vec{x}, \vec{\xi}) = (2\pi^2 a)^{-\frac{1}{2}} \exp[-\frac{1}{2a^2}(\vec{X} - \vec{\xi})^2]$$

where a is a constant having dimensions of length. then a physical quantity $F(\xi, \eta, \zeta)$ in the abstract space can be represented in the physical space as

$$\tilde{F}(x, y, z) = \int\int\int_{-\infty}^{+\infty} F(\xi, \eta, \zeta)\chi(\vec{X}, \vec{\xi})d\xi d\eta d\zeta \qquad (4.44)$$

F may depend on t also and hence the $\tilde{F}$. One of the possible consequences of this approach is that point do not exist in the physical space. They have their counterparts in charge proportional to the functions $\chi(\vec{X}, \vec{\xi})$ as considered above. Such a charge distribution has a finite (and positive) electrostatic energy and, at the same time, is stable since it can not be seperated into smaller parts. It appears that the above formation is suitable for classical theory and is not suitable for quantum theory. It is possible to extend the above formalism so as to fit them in the formalism of quantum theory. For this let us first consider the three dimensional space, in which we have the probability density function χ as given above. Now according to quantum theory, the probability amplitude or wave function Φ may be written as

$$\Phi(\vec{X},\vec{\xi}) = (2\pi a^2)^{-\frac{1}{2}} \exp[-\frac{1}{\pi^2}(\vec{X}-\vec{\xi})^2] \tag{4.45}$$

Now the question is : **how can we understand the existence of this wave function on the basis of quantum theory?**

In Schrödinger framework, a coordinate x and a momentum $\hat{p}$ are associated with a particle and as such no mass of the particle enters into the definition of these operators. Therefore, they can be applied to the case of a particle having no mass or even to the case of a mathematical point rather than a physical particle. Thus we arrive at the idea that the points of space should be associated with them not only by coordinates but also by momenta. Born used this idea to a certain extent in his reciprocity theory [Prugovecki, 1984] where the coordinates and momenta of a particle enter in a symmetrical manner. We have already discussed Yukawa's concept of non-local fields where the field variables are considered as functions of position and momentum variables.

Let us describe the points in a given portion of a space using the coordinates $\hat{X} = (\hat{x}_1, \hat{y}_2, \hat{x}_3)$ and momenta $\hat{P} = (\hat{p}_1, \hat{p}_2, \hat{p}_3)$. Here, $\hat{X}$ and $\hat{P}$ satisfies the usual commutation relation. Now a complex variable or operator can be constructed out of $\hat{x}$ and $\hat{p}$ as

$$\hat{\xi} = \hat{x} + i\alpha\hat{p}$$

where α is a dimensional constant.

In the case where the variable $\hat{\xi}$ is known exactly, can be described by

$$\hat{\xi}\Phi = \xi\Phi$$

In the Schrödinger representation this becomes

$$\left(x_k + \alpha\hbar\frac{\partial}{\partial x_k}\right)\Phi = \xi_k\Phi \tag{4.46}$$

Taking $a^2 = \frac{1}{2}\alpha\hbar$, we obtain the solution as

$$\Phi = A\exp\left[-\frac{(\vec{X}-\vec{\xi})^2}{4a^2}\right] \tag{4.47}$$

where A is a constant . Similarly we can get a solution for Φ^* as

$$\Phi^* = A^* \exp[-\frac{(\vec{X}-\vec{\xi})^2}{4a^2}] \tag{4.48}$$

Taking $\hat{\xi}^* = \hat{x} - i\alpha\hat{p}$ now we can write

$$\begin{aligned} \langle \vec{X} \mid \vec{\xi} \rangle &= A \exp[-\frac{(\vec{X}-\vec{\xi})^2}{4a^2}] \\ \langle \vec{\xi}^* \mid \vec{X} \rangle &= A^* \exp[-\frac{(\vec{X}-\xi)^2}{4a^2}] \end{aligned} \tag{4.49}$$

Using the commutation relations between coordinates and momenta we obtain the following commutation relations

$$\hat{\xi}_j\hat{\xi}_k^* - \hat{\xi}_k^*\xi_j = 4a^2\delta_{ij} \tag{4.50}$$

This implies that both $\hat{\xi}$ and $\hat{\xi}^*$ can not be determined sharply in the same state. The elements of space can be labeled by the eigenvalues $\hat{\xi}$ or $\hat{\xi}^*$. It will be assumed that physical space is characterized by the fact that $\hat{\xi}_k$ and $\hat{\xi}_k^*$ are real. Taking the value of $A \simeq (2\pi a^2)^{-\frac{3}{4}}$,

$$\langle \vec{\xi}^{*\prime} \mid \vec{\xi}^{\prime\prime} \rangle = \exp[-\frac{(\vec{\xi}^{*\prime} - \vec{\xi}^{\prime\prime})^2}{8a^2}] \tag{4.51}$$

$$\langle \vec{\xi}^{*\prime} \mid \vec{X} \mid \vec{\xi}^{\prime\prime} \rangle = \frac{1}{2}(\vec{\xi}^{*\prime} + \vec{\xi}^{\prime\prime}) \exp[-\frac{(\vec{\xi}^{*\prime} - \vec{\xi}^{\prime\prime})^2}{8a^2}] \tag{4.52}$$

For $\vec{\xi}^* = \vec{\xi}$, it follows that

$$\langle \vec{\xi}^* \mid \vec{\xi} \rangle = 1; \qquad \langle \vec{\xi}^* \mid \vec{X} \mid \vec{\xi} \rangle = \xi^*$$

This formalism leads to the possibility of non-local fields. Suppose a field is described by a Hermitian function operator $\hat{F}$. Now, if it has exact values in the ξ, ξ^* space, we can write as

$$\hat{\xi}\hat{F} - \hat{F}\hat{\xi}^* = 0 \tag{4.53}$$

We shall call this variable satisfying the above relation as a "basic variable". In the mixed representation, the above relation can be expressed as

$$(\vec{\xi}^{\prime} - \vec{\xi}^{*\prime\prime})\langle \vec{\xi}^{\prime} \mid \hat{F} \mid \vec{\xi}^{*\prime\prime} \rangle = 0 \tag{4.54}$$

or,

$$\langle \xi' \mid \hat{F} \mid \vec{\xi}^{*\prime\prime} \rangle = F(\vec{\xi}')\delta(\vec{\xi}' - \vec{\xi}^{*\prime\prime}) \tag{4.55}$$

In the X-representation, we have

$$\langle \vec{x}' \mid \hat{F} \mid \vec{x}'' \rangle = \hat{F}(\frac{\vec{x}' + \vec{x}''}{2}) \exp\left[-\frac{(\vec{X}' - \vec{X}'')^2}{8a^2}\right] \tag{4.56}$$

where

$$\hat{F}(\vec{x}) = (2\pi a^2)^{-\frac{3}{2}} \int F(\vec{\xi}) \exp[-\frac{(\vec{\xi} - \vec{X})^2}{2a^2}] d^3\xi$$

Expectation value is given by

$$\langle \vec{X} \mid \hat{F} \mid \vec{X} \rangle = \hat{F}(\vec{X}); \quad \text{where} \quad \hat{F}(\vec{X}) = \int F(\vec{\xi})\kappa(\vec{X}, \vec{\xi}) d^3\xi$$

Let us suppose that we have two functions $\hat{F}$ and $\hat{G}$. Then we have the following relation for their product as

$$\begin{aligned} \langle \vec{X} \mid \hat{F}\hat{G} \mid \vec{X}' \rangle &= \int \hat{F}\frac{(\vec{X}' + \vec{X}'')}{2} \hat{G}\frac{(\vec{X}' + \vec{X}'')}{2} \\ & \exp\{-\frac{1}{8a^2}\left[(\vec{X} - \vec{X}'')^2 + (\vec{X}' - \vec{X}'')^2\right]\} d^3x'' \end{aligned} \tag{4.57}$$

However, in the mixed representation, we have

$$\langle \vec{\xi} \mid \hat{F}\hat{G} \mid \vec{\xi}^{*\prime} \rangle = F(\vec{\xi})G(\vec{\xi}^{*\prime} \exp\left[-\frac{(\vec{\xi} - \vec{\xi}^{*\prime})^2 + (\vec{X}' - \vec{X}^{*\prime})^2}{8a^2}\right] d^3x'' \tag{4.58}$$

Again

$$\langle \vec{X} \mid \hat{F}\hat{G} \mid \vec{X}' \rangle = \int \hat{F}\frac{(\vec{X} + \vec{X}')}{2} \hat{G}\frac{(\vec{X} + \vec{X}')}{2} \exp[-\frac{(\vec{X} - \vec{X}')^2}{4a^2}] d^3x' \tag{4.59}$$

and also

$$\int \langle \vec{X} \mid \hat{F}\hat{G} \mid \vec{X}' \rangle d^3x = 8\pi^{\frac{3}{2}} a^2 \int \hat{F}(\vec{X})\hat{G}(x) d^3x \tag{4.60}$$

So far we have consider the three dimensional problems. Let us now take the case of four dimensional space-time as

$$x^0 = ct,\ x' = x,\ x^2 = y,\ x^3 = z$$

and the metric tensor $g_{\mu\nu} = g^{\mu\nu}$ with diagonal elements $+1, -1, -1, 1$, and then we can write

$$\hat{\xi}^\mu = \hat{x}^\mu - i\alpha g^{\mu\nu}\hat{p}_\nu$$

Then we obtain the solution as

$$\Phi = B\exp\{[(x^0 - \xi^0)^2 - (\vec{X} - \vec{\xi})^2\} \tag{4.61}$$

with B as a constant.

It should be noted that this solution can not be used for all space-time since it goes to infinity if $| x^0 | \to \infty$. Therefore, we take the space-like hypersurface characterized by a normal vector n^μ,

$$n_\mu n^\mu = \text{constant}; \quad n_\mu n^\mu = 1, \quad n_0 > 0$$

and consider only values of x^μ and ξ^μ corresponding to points on this hypersurface. Thus we have

$$n_\mu(x^\mu - \xi^\mu) = 0; \quad \text{and} \quad x^0 - \xi^0 = -\frac{1}{n_0} n_\mu (x^k - \xi^k) \tag{4.62}$$

So, Φ becomes

$$\phi = B\exp[\frac{1}{4a^2}(\frac{n_i n_j}{n_0^2} - \delta_{ij})(x^j - \xi^j)(x^k - \xi^k)] \tag{4.63}$$

This remains finite everywhere on the hyperplane. It is well known that the line element on the hyperplane can be written as

$$\begin{aligned} ds^2 &= r_{ik} dx^j dx^k \\ r_{jk} &= -\delta_{jk} + \frac{n_j n_k}{n_0^2} \\ r &= -\frac{1}{n_0^2} \end{aligned} \tag{4.64}$$

Here r is the determinant of r_{jk}. So, in the intergration over the hyperplane, the invariant volume element can be taken as $n_0^{-1} d^3s$. Now by means of a

Lorentz transformation, we can go over to a frame of reference such that the hyperplane becomes simply the three-dimensional space at a constant time ($n_0 = 1$, $n_k = 0$). This framework may give a new formalism of non-local fields. The basic element is a Gaussian distribution function (rather than point), and this can be related with the elementary particles. This approach may help one to avoid infinities that the point model leads to.

4.4. Probabilistic Finsler Geometry :

In the previous section **4.2**, we have considered the fluctuation in the extra dimensions of 4-dimensional Minkowski metric. So this is a kind of probabilistic approach to Kaluza-Klein theory. But when the Kaluza-Klein ansatz of multidimensionality is taken seriously with the purpose of constructing realistic models of physical field interactions, the investigators are faced with many difficulties, of which the problems of cosmological constant, of light fermions and of gauge hierarchy are regarded as greatest nowadays. Asanov [Asanov 1988] tried to look for the true multidimensionellness in the nearest vector fibrations over space-time.

To defend this approach, we must say that, as compared with the narrow sense treatment of the space-time as merely a set of points x^i, the concepts of fibration of tangent vectors and, then, of higher-degree tangent vectors reflect the real physical images more adequately. The real physical images involve the velocities and accelerations as quite self-contained and independent entities complementary to point x_i. In fact, the velocity and acceleration concepts underlie the statistical pattern of space-time. We have developed Finslerian gauge ideas to offer a consistent alternative to conventional Kaluza-Klein approach. The basic motivation is that instead of seeking gauge-covariant fibered solutions of the equations formulated on amorphous (read "non-fibered") multidimensional spaces, and, then reduced by imposing the conditions of cylindricity in a genuine artificial way, it looks more expedient, de rerum naturae, to immediately formulate generalized gauge equations on suitable fibrations over space-time manifold. The two approaches do not commute with each other.

In conformity with our basic approach towards stochastic geometry, we shall discuss our theory to the homogeneous case, that is when the basic geometric object possess the property of zero-degree homogeneity with respect to each vector of the sequence $(y, a, \ldots)$. This restriction is in correspondence with the viewpoint that the theory should involve dependence only on the directions of vectors, but not on their lengths. For example, the real physical three dimensional velocity (three components of four vector over its fourth component) is entirely given by the direction of four vector. This example arguing for usage of the homogeneous case only, may be extended to the vectors of higher orders of tangency. If the rule stating that all the physical phenomena depend only on direction of four-dimensional vectors is universal, then only the homogeneous case of our theory is adequate to the physical pattern of the world.

Therefore, in making various calculations, we go over conveniently to independent parameters $z, u, \ldots$ which are genuine arguments of the homogeneous approach. The general dimensionality $4+4+4+\ldots$ of the total vector fibration over space-time reduces, then to $4+3+3+\ldots$. The total set of coordinate and gauge transformations used in the parametrical representation of the homogeneous theory reads

$$\begin{aligned} x^i &= x^i(\overline{x}) \\ Z^P &= Z^P(x, z') \\ u^A &= u^A(x, z, u^*) \end{aligned} \tag{4.65}$$

which is distinguishable by its triangular structure from Kaluza-Klein's ansatz $X^V = (X^V(\overline{X}^U)$, where, $U, V = (i, P, A, \ldots)$. The $(4+3+3+\ldots)$-fold variational principle is intrinsic to the homogeneous theory. The multidimensional structure impel one to invoke the idea to average it over all the "internal" variable z and u to make comparison with observable entities. Apparently, the idea of averaging over the tangent and higher-order vector must be regarded as a principle of general significance underlying the stochastic theory of space-time, on the one hand and probably the nonlocal quantum field theory, on the otherhand. In its rigorous sense, the principles of gauge-covariant averaging should read as follows :

Given any object with the indices of pure external type, for example, the basic metric tensor g_{ij}, $(x, z, u, \ldots)$, the base curvature tensor $L^n_{kij}(x, z, u, \ldots)$, a mixed tensor $w^j_i(x, z, u, \ldots)$ etc. Using the total Jacobian $G(x, z, u, \ldots)$, we get,

$$a_{ij}(x) = < g_{ij} > \stackrel{def}{=} \frac{\int g_{ij} G dz^{N-1} du^{N-1} \ldots}{\int G dz^{N-1} du^{N-1} \ldots} \tag{4.66}$$

and literally the same definitions for $< L^n_{kij} >$ and $< w^j_i >$. In otherwords, the total Jacobian G plays the role of the distribution function with respect to the "internal" variables $z, u, \ldots$.

It should be stressed that the above procedure yields the result of scalar nature under the gauge transformations and also retains the tensor properties of averaged objects under the general coordinate transformation of the base points x. Treat the averaged entities as observables in a conventional sense and consider the differences including

$$e_{ij} \stackrel{def}{=} g_{ij} - < g_{ij} > \tag{4.67}$$

as small fluctuations for the purpose of making estimations.

It is worth mentioning that the dependence of g_{ij} on $z, u, \ldots$ may be interpreted as a manifestation of the existence of some internal curvature. Be this curvature entirely negligible, we would get merely $g_{ij} =< g_{ij} >$, so that we may conclude that the smallness of fluctuations must be determined by the smallness of the internal curvature. By averaging the unit tangent vector proper, we get the vector field

$$U^q(x) \stackrel{def}{=} \frac{\int l^q G dz^{N-1} du^{N-1} \ldots}{G dz^{N-1} du^{N-1} \ldots} \tag{4.68}$$

which would be of interest to treat in the context of the modern aether theories of space-time, as well as in the context of the modern general relativistic frameworks admitting the presence of the preferred reference frames in the Universe.

So, figuratively speaking, we have $(4+3+3+\ldots)$-dimensional probabilistic distributions over the four-dimensional space-time manifold. In case of the proper Finsler geometry, that is, over $(4+3)$-dimensional probabilistic distributions, we can get conventional $(4+1)$-dimensional Klein-Kaluza framework for a preferred frame of reference. Then, by compactification of this extra dimension, it is possible to get a deeper insight to the origin of quantum laws. It also raises new possibility in studying the multidimensional Finslerian fields and distribution theory which might have deeper significance in cosmology.

4.5. Local Heat Energy and Local Temperature :

It is well known that the heat energy (Q), temperature (T) and entropy (S) correspond to some form of internal motions. But the problem lies in thier definitions with respect to these motions. The development of relativistic thermodynamics and quantum theory gives rise to new understanding of these quantities and new debate has been started.

The controversy has its roots in two evident remarks by Eddington [Eddington 1923] : "Since one can obtain energy from heat it is not evident that they transform in a different way under a Lorentz transformation when $\beta \to 1$" or if heat describe internal energy associated with internal motions in one inertial frame , one expects it would also transform like energy in a Lorentz transformation".

Recently Fenech and Vigier [Fenech 1996] tried to give a precise relativistic definition of Q and T from which one can derive directly the form of their variation under Lorentz transformations. This justifies Planck's initial assumptions. This approach is very much similar to the idea of introducing the internal variable to describe the extended object with unsharp boundary or hazy lump. We have already shown in section **4.3** that these internal variables may be of random in nature. With Einstein's definition of a non-homogeneous body in calorific equilibrium in mind, one can define such a body by the existence of an inertial frame R_0 in which the total volume V_0 is constant. Now this volume can be sudivided in a set of equal volumes or droplets $d\tau_0$ which contain the same energy w_0 and a total momentum $P_0 = 0$ called Einstein's boxes (E-box). Then in another frame moving with a velocity v, we can write

$$w = (w_0 + vP_0)(1-\beta^2)^{-1/2} \tag{4.69}$$

The total energy of the hot body is thus given by the relation

$$W = \sum w = W_0(1-\beta^2)^{-1/2}$$

In this case, W_0 denotes the total energy of the body in R_0. Again, by definition

$$W_0 = M_0c^2$$

i.e., one can attribute a proper rest mass M_0 to this hot body. So,

$$W = W_0(1-\beta^2)^{-1/2} = M_0c^2(1-\beta^2)^{-1/2} \tag{4.70}$$

Then if we consider M_0c^2 as the quantity of heat Q_0 in V_0, we can write

$$Q_0 = M_0c^2, \quad W = \frac{Q_0}{\sqrt{(1-\beta^2)}}. \tag{4.71}$$

This is in agreement with relativity theory. At this point, Fenech and Vigier [Fenech 1996] considered the supplementary assumptions i.e., each E-box is filled with identical constitutive elements (particles, molecules etc.) with identical rest masses m_0 (endowed with a random internal distribution of velocities) which behave like moving δ-functions. But this amounts to introducing in each $d\tau_0$:

(1) a conserved internal energy-momentum i.e., $\partial^\nu t_{\mu\nu} = 0$, and,

(2) a conserved current density distribution i.e.,$\partial_\mu j_\mu = 0$.

In fact, this is the starting point of Bohm and Vigier's [Bohm 1954] first analysis of relativistic extended body. Here, the individual element $\{j\}$ which moves in a random way along individual stochastic paths, can pass from one $d\tau_0$ to another. But in local equilibrium situation, it is assumed to be constant in any $d\tau_0$. Bohm and Vigier [Bohm 1954] showed that the existence of conserved $t_{\mu\nu}$ and J_μ in an E-box implies that one can associate with each $d\tau_0$ two remarkable average points Y_μ and X_μ as basis for the definition of collective variables. These two points can be defined in the following way :

(1) A center of gravity Y_μ defined in P_0 by the relations

$$Y_\mu \in t_\alpha dv = \int x_\mu t_\alpha dv$$

.

This point moves with a four velocity W_μ ($W_\mu W_u = -c^2$) parallel to P_μ and one can define an angular momentum $M_{\mu\nu}$ and a spin vector S_μ with respect to Y_μ by the relations

$$\begin{aligned} M_{\mu\nu} &= \int [(x_\mu - Y_\mu)t_\nu - (x_\nu - Y_\nu)t_\mu]dv \\ S_\mu &= (1/c)\overline{M}_{\mu\nu}W^\nu \end{aligned} \tag{4.72}$$

where

$$\overline{M}_{\mu\nu} = \frac{1}{2}\epsilon_{\mu\nu\alpha\beta}M^{\alpha\beta} \tag{4.73}$$

Let us consider the proper time θ along the Y_μ world line (i.e., $\dot{A} = \frac{dA}{d\theta}$). In the absence of exterior forces, one can write $\dot{P}_\mu = 0$, so that writing $P_\mu P^\mu = -M_0^2c^2$, we have

$$P_\mu = MW_\mu$$

This also implies that $\dot{S}_\mu = 0$. Then,

(2) A center of mass density X_μ is defined in S_0^1 which is a particular type of inertial frame (Möller frame) in which $P_i = 0$, $i = 1, 2, 3.$ and it is given by the relation

$$X_\mu \int j_0 dv = \int x_\mu j_0 dv \tag{4.74}$$

Here, X_μ follows the world line associated to a proper time τ and four velocity

$$\dot{X}_\mu = U_\mu \quad (\text{with } \dot{X}_\mu \dot{X}^\mu = -c^2), \tag{4.75}$$

so that $\dot{X} = \frac{dX}{d\tau}$ represents the average derivatives along the X_μ path. In this framework, an intrinsic angular momentum in $d\tau_0$ can be defined as

$$\begin{aligned} S_{\mu\nu} &= \int [(x_\mu - X_\mu)t_{\nu_0} - [x_\nu - X_\nu)t_{\mu_0}]dv \\ &= M_{\mu\nu} + R_\nu P_\mu - R_\mu P_\nu \end{aligned} \tag{4.76}$$

This implies that the total angular momentum is the sum of an intrinsic $S_{\mu\nu}$ and orbital $(R_\nu P_\mu - R_\nu P_\mu)$ angular momentum. Then one can define

$$R_\mu = X_\mu - Y_\mu = (\frac{(1}{M_0 c^2})S_{\mu\nu}P^\nu \tag{4.77}$$

which satisfies the following relations for free particles i.e.,

$$\dot{P}_\mu = 0, \quad M_0^2 c^2 R_\nu = -S_{\mu\nu}P^\mu \tag{4.78}$$

$$\dot{S}_{\mu\nu} = P_\mu \dot{X}_\nu - P_\nu \dot{X}_\mu \tag{4.79}$$

$$P^\mu R_\nu = 0 \tag{4.80}$$

It yields the relation $\frac{d\tau}{d\theta} = \frac{M_0}{m_0}$, when one introduces the inertial mass m_0' by the relation

$$P_\mu \dot{X}^\mu = -m_0' c^2$$

The bilocal system $(Y_\mu(\theta), X_\mu(\tau))$ behaves like a microscopic flat clock where moving needle R_μ, centered on Y_μ, rotates in a plane orthogonal to P_μ (since $P_\mu R_\mu = 0$).

Now with the help of these collective variables $Y_\mu(\theta)$ and $X_\mu(\tau)$, it is possible to study the internal motions and the corresponding thermodynamic entities. It is assumed that in a local E-box, the center of mass and center of matter density coincide on the average for all practical purposes (FAPP) in a small proper time i.e., $\overline{Y}_\mu^\beta = \overline{X}_\mu^\beta$. Here, the bar denotes average values taken over a small proper time interval $d\theta$ associated with Y_μ : a property which does not exclude small relative oscillation within $d\theta$. Following, this procedure Fenech et al. [Fenech 1996] suggested two important facts :

(a) The quantity of heat Q_0 of an Einstein Box (E-box) should be defined as the total energy $E_0 = Q_0 = M_0 c^2$ associated with the isotropic spherically symmetric stochastic internal set of motions of its constitutive elements around its center of mass Y_μ' in its rest frame S_0^1.

(b) The heat Q observed in any other inertial frame S is the energy associated in S with the isotropic distribution (around Y_μ) of the internal motions in the said S-flow.

But according to relativity theory, Q_0 does not transform into Q like energy under a Lorentz transformation. In this framework, each of the two pair elements in S_0^1 has a different probability weight with respect to an isotropic distribution in different way to the heat energy. This restriction of heat energy to the energy of spherical internal isotropy, thus, explains why one can not transform Q according to the usual Lorentz transformation. However, this difficulty can be tackled if we consider the local definitions of Q and T. They are as follows :

The local heat energy Q_0 of an E-box is related to its isotropic internal random motions around a center of mass Y_μ in its inertial rest frame S_0^1 and the heat Q observed in any other inertial frame S corresponds to internal isotropic motions around Y_μ. Again the internal chaotic motion of each E-box can be represented by a temperature T_0 which corresponds to the average kinetic energy of their center of matter density X_μ which spirals around Y_μ. The local definitions imply the existence of two currents :

(1) a current of heat which follows the path of Y_μ;

(2) a current of temperature defined by the X_μ motions.

With these local definitions one can get the corresponding laws of transformation i.e.,

$$\begin{aligned} Q &= Q_0(1-\beta^2)^{\frac{1}{2}} \\ T &= T_0(1-\beta^2)^{\frac{1}{2}} \end{aligned} \tag{4.81}$$

Now, if we look into our general notion of hazy lumps (i.e., C-spaces) to build up stochastic metric, where those leptons are considered as spherically symmetric, then it might be possible to assign small non-zero temperature to the vacuum itself. In that case, the **2.7°K** cosmological background radiation might be related to the fluctuation of the space-time metric. It may have significant role also in noise induced reduction of wave packet [Stapp 1992].

ASTRO-PARTICLE PHYSICS

"The fountains mingle in the river
And the rivers with Ocean,
The winds of heaven mix for ever
With a sweet emotion.
Nothing in the world is single,
All things by a law divine
In one spirit mix and mingle."

Percy B. Shelley : Love's Philosophy.

5. ASTRO-PARTICLE PHYSICS

The fluctuation of space-time may be detected at small distances. Namsrai [Namsrai 1996] analysed the experimental data for scattering and got a limit of the fundamental length the length scale of fluctuation, as of the order of 10^{-31} cm i e., larger than Planck length. The stochastic nature of space-time can be revealed at ultra high energy interactions of elementary particles. The spectrum of highest energy cosmic ray primary proton is one of the challenging problems in astroparticle physics. In a recent work it has been shown [Roy 1995] that frame dependence of the scattering matrix (which is related to the fluctuation of space-time) shed new light on this problem.

The acceleration mechanism of cosmic ray particles of energies from a few Mev to enormous values of about 10^{21} eV is another big issue in high energy Astrophysics. We shall also discuss this problem in this chapter in relation to fluctuation of vacuum.

5.1. Acceleration Mechanism

Jokipii [Jokipii 1979] made a simple and comprehensive classification of acceleration mechanism in the form of a two by two matrix with four entities as follows :

	Continued or closed	Extended or open
Causal or Deterministic	SWANN MECHANISM DOUBLE LAYERS E_{11}	SHOCK WAVES
Stochastic	No known confined Stochastic Mechanism	Fermi and Descendants ZPF Acceleration Mechanism

According to Jokipii, Acceleration Mechanism (AM) can be represented in two major and independent classification schemes :

- **Deterministic versus stochastic**
- **Confined versus extended**

Acceleration by the rotating field of a pulsar is an example of a both deterministic and confined AM [Thielheim 1975]. Acceleration by shock waves is extended and deterministic [Völk 1987]. The Fermi [Fermi 1949] AM is an example of an extended stochastic AM. The Fermi mechanism has had many variants and descendants. The AM with random zeropoint field [Rueda 1990] is also an example of an extended stochastic mechanism which is a remote relative of the Fermi one. However, it distinguishes itself from the Fermi AM and its near relatives. It is not being directly and fully dependent on plasma physics considerations. Extended AM are also called diffuse acceleration until recently, the possibilities of success for diffuse AM seemed however, severely limited by theoretical difficulties associated with simultaneous acceleration and propagation in interstellar space ([Cowsik 1979, 1980, 1985], Hayakawa [1969]; Giler et al. [Giler 1985]). Lerche and Schlickeiser [Schlickeiser 1989] proposed a stochastic acceleration where acceleration and propagation may occur simultaneously. They explained this by means of a reanalysis of the usual transport equation in the Leaky box model. Exact analytical solutions of the transport equations in the steady state and time dependent case are found and compared with observations. The solutions indicate the coexistence of monoenergetic, power law, exponential and relativistic Maxwellian particle spectra. Since these types of spectra are observed in cosmic sources, the results of Schlickeiser et al.

[Schlickeiser 1981] are in favor of continuous acceleration models. The secondary / primary ratio in galactic cosmic radiation is calculated in this framework and the validity of Cowsik's [Cowsik 1980] and Eichler's [Eichler 1980] argument against continuous distributed acceleration has been questioned. According to Schlickeiser argument if it is assumed that the acceleration of cosmic rays takes place mostly in a low density halo or our Galaxy, whereas the secondary one produced in the dense interstellar clouds confined to a thin layer in the galactic plane, then, due to the special position of the Solar system near the production region of the secondaries, the observed difference between the energy spectra of primaries and secondaries could be explained as a local phenomena. However, [Völk, 1987] there are some difficulties with such a picture. One can think of a continuous shock acceleration but it requires a very large energy input to accelerate all the primaries which escape before arriving on the disk.

On the otherhand, the shock wave acceleration mechanism as advocated by Völk et al. [1987] has drawn large attention to understand the particle acceleration at high energy. This framework has been shown to be very useful up to 10^{15} eV. But the spectra of ultra high energy cosmic ray particle up to 10^{19} eV or above possesses one of the outstanding problems within this framework of diffuse shock acceleration [Blandford]. Besides the shock wave acceleration, there are other theories which appear to be also promising to explain the energy spectra of ultrahigh energy particles. Another mechanism of particle acceleration has been proposed using the idea of magnetic reconnection. Here, the particle energy gain is directly proportional to the product of the pre-emerging magnetic field and the plasma resistivity. In general cosmic plasma have electrical conductivities and magnetic Reynolds' number large enough to keep the field like "frozen into" the medium. Particle acceleration to relativistic energies is possible provided that plasma turbulence or some form of instability leads to a drastic reduction of the classical conductivity. It is perhaps premature to determine whether the magnetic reconnection may be considered as alternative to shock-wave acceleration mechanism.

In this chapter we shall discuss the main ideas behind the stochastic acceleration, acceleration mechanism induced by the Zero Point Field (ZPF) and finally due to the fluctuation of space-time metric. We have already pointed out that Fermi [Fermi, 1943] proposed the ingradients of present stochastic acceleration mechanism. Instead of particles gaining energy in one giant steps, same acceleration mechanism postulates that wave turbulence causes particles to diffuse through momentum space in many small steps Δp. Let Δt be the time interval between these steps. we can define a diffusion coefficient

$$D_{pp} = \langle \frac{\Delta\vec{p}\Delta\vec{p}}{\Delta t} \rangle \tag{5.1}$$

and the distribution function $f(\vec{p}, \vec{n}, t)$ satisfies the following Fokker-Planck equation

$$\frac{\partial f}{\partial t} = \nabla_p . D_{pp} \, \nabla_p \, f \tag{5.2}$$

In case of Alfven waves, it interacts resonantly with particles whose Larmor radii r_L match their wavelength. Now if we think of Alfven waves as quanta with momenta $\hbar\vec{k}$ and energy $\hbar\omega$, then the ratio of their energies to their momenta is much smaller than the corresponding ratio for high energy particles, which is of the order of c. Therefore, Alfven waves are most effecient at adding momentum in steps $\Delta\vec{p}$ almost perpendicular to $\vec{p}$, that is to say Alfven waves are more likely to be involved in scattering than directly accelerating particles. The scattering rate is usually expressed as a diffusion co-efficient in the pitch angle ϕ, as

$$D_{\phi\phi} \equiv \langle \frac{\Delta\phi^2}{\Delta t} \rangle \approx (\frac{\delta B}{B})^2_{k \sim \gamma_L^{-1}} (\frac{c}{\gamma_L}) \tag{5.3}$$

where $(\frac{\delta B}{B})$ is the relative amplitude of the resonant Alfven waves. Again the spatial diffusion coefficient associated in the scattering is written as

$$D_{nn} \equiv \langle \frac{\Delta x^2}{\Delta t} \rangle \sim \frac{v^2}{v} \tag{5.4}$$

Magnetosonic waves are more effective at accelerating particles. Here the fundamental resonance is that associated with Landau damping $\omega = k_{||}v_{||}$, where $||$ denotes the projection along the magnetostatic field and ω is the magnetosonic wave frequency. In this cases, energy diffusion coefficient becomes

$$D_{EE} \sim E^2(\frac{a}{v}) \, \omega \, (\frac{\delta B}{B})^2 \tag{5.5}$$

This type of acceleration can also occur in non-linear Landau damping of the beat wave formed by two Alfven waves. In that case of relativistic particles, the rate of energy gain is proportional to the energy as in Fermi process. In order to produce a power law distribution function it requires this prescription for acceleration to be combined with an exponential distribution of acceleration times. Process like this may be responsible for creating the seed particles for more effecient acceleration process. However, in the stochastic acceleration mechanism by wave turbulence the quasi-linear approach underlying Fokker-Planck formalism is questionable. It may be mentioned that the newly discovered Cosmic Ray (CR) components called "anomalous" CR and their properties measured in the Indian CR experiment "ANURADHA" in spacelab 3 indicate that the basic acceleration process may be of stochastic in nature. We have already mentioned that acceleration mechanism with random zero point field (ZPF) is a variant of stochastic mechanism and relative of Fermi one.

Now, let us consider briefly the acceleration mechanism within this approach. At first we shall consider the acceleration mechanism (AM) for poarizable particle in ZPF and then for monopolar particle.Let us start with Einstein and Hopf model.

Einstein and Hopf Model :

The one dimensional case will be first considered as it is popolurly known as Einstein-Hopf model (EHM). Here, the translational motion of particle is considered along one single axis, say, x-axis and to vibrate parallel to an orthogonal axis, say, the z-axis. For this we shall calculate the energy growth under ZPF. Let us start with a nonrelativistic massive monopolar particle where oscillating dipole moment $p = zql_z$ is along the parallel to the Z-axis. The Abraham-Lorentz equation (ALE) for the oscillating dipole is then

$$\ddot{p} - \Gamma_q \dddot{p} + \omega_0^2 p = \frac{2q^2}{3\mu} E_z \tag{5.6}$$

where $\Gamma_q = \frac{2q^2}{3\mu c^3}$ is the Abraham-Lorentz time parameter of radiation damping, q is the oscillating charge and μ the mass associated with the charge. Here. μ is not the mass of the particle but it is the mass of the oscillating subparticle. The inertia from the mass of the particle, m, only affects translational motions. ω_0 is the charachteristic frequency of the oscillator that in a first approximation is assumed to be linear. E_z is the Z-component of the electric field $\vec{E}$ of the random radiation. Effects due to $\vec{B}$ field are negligible in the nonrelativistic approximation. In this framework, the random electromagnetic field is written with a random phase angle in the spirit of Planck and Einstein.

$$E(x,t) = R_l \sum_{\lambda 01}^{2} \int d^3k \hat{\epsilon}(\bar{k},\lambda)\eta(\omega_k) \exp[i\omega t - i\bar{k}.\bar{x} - i\theta(\bar{k},\lambda)] \tag{5.7}$$

Where,

$$\hat{\epsilon}(\bar{k},\lambda).\bar{k} = 0; \quad \hat{\epsilon}(\bar{k},\lambda).\hat{\epsilon}(\bar{k},\lambda') = \delta_{\lambda\lambda'} \tag{5.8}$$

For the case each wave vector $\bar{k}$ then there exists two directions of polarization $\hat{\epsilon}(\bar{k},\lambda), \lambda = 1, 2,$ representing the two mutually orthogonal polarization vectors. The fluctuating field has a spectral energy density ρ which for thermal case can be written as

$$\begin{aligned} \rho(\omega,T)d\omega &= \frac{\hbar\omega^3}{2\pi^2c^3}\text{cotgh}\frac{\hbar\omega}{2kT}d\omega \\ &= \frac{\hbar\omega^3}{\pi^2c^3}\left[\frac{1}{\exp(\frac{\hbar\omega}{kT})-1}\right] \end{aligned} \tag{5.9}$$

Now for ZPF,ρ is expressed as

$$\rho(\omega,0)d\omega \equiv \rho_0(\omega)d\omega = \frac{\hbar\omega^3}{2\pi^2c^3}$$

ρ is quadratically proportional to $\eta(\omega_{\bar{k}})$ i.e.,

$$\rho = \frac{\omega^2}{c^3}\eta^2(\omega)$$

The random phase angle $\theta(\bar{k},\lambda)$ is a different random variable for each $(\bar{k},\lambda)$ pair. It is uniformly distributed between 0 and 2π. The corresponding magnetic field is

$$B(x,t) = \text{Re}\sum_{\lambda=1}^{2}\int d^3k[\hat{k}\times\hat{\epsilon}(\bar{k},\lambda)\eta(\omega_{\bar{k}})\exp[i\omega_{\bar{k}}t - i\bar{k}.x - i\theta(\bar{k},\lambda)] \tag{5.10}$$

where $\hat{k} = \bar{k}/k,\ k = \mid \bar{k} \mid$. When ALE is written for ZPF + thermal radiation in SED, it may be called as Braffort-Marshall equation [BME]. The BME for p of the above equation (5.6) can be solved

$$\begin{aligned} p &= \sum_{\lambda=1}^{2}\int d^3k\frac{3c^3}{2\omega^3}\epsilon_z(\bar{k},\lambda)\eta(\omega_{\bar{k}})\sin\alpha(\omega_{\bar{k}}) \\ &\times\ \cos(\omega_{\bar{k}}t - \bar{k}.x - \alpha(\omega_{\bar{k}}) - \theta(\bar{k},\lambda)) \end{aligned} \tag{5.11}$$

with $tg\alpha(\omega) = \frac{\Gamma_q\omega^3}{\omega_0^2-\omega^2}$. For simplicity, we have written ω for $\omega_{\bar{k}} = c \mid \bar{k} \mid$.

Uptill now we have considered the vibrational motion. Now let us proceed with the translational motion along x-axis (i.e, the axis along which the particle is allowed to displace). The force in the x-direction can be written as

$$F_x = p\frac{\partial E_x}{\partial z} - \frac{1}{c}\dot{p}B_y \tag{5.12}$$

where the limit when the dipole length goes to zero is implied in the first term. The force acts in such a way that, if at time t the translational velocity along the x-axis of the massive particle is v_t, at time $t+\tau$, it is $v_t+\tau$:

$$mv_{t+\tau} = mv_t + \Delta_\tau \tag{5.13}$$

m being the mass of the particle and Δ_τ the impulse . The impulse is given by by

$$\Delta_\tau = \int_t^{t+\tau} F_x dt$$

Using Maxwell equation we obtain

$$\Delta_\tau = \int_0^\tau p\frac{\partial E_z}{\partial x}dt \tag{5.14}$$

This term is usually neglected on the grounds that the term is independent of τ. Now $< v_t\Delta_\tau >= 0$ as the velocity at time t, is v_t and the impulse due to the incoming ZPF at later time $(t, t+\tau)$ can not be correlated,

$$< (mv_t+\tau)^2 >=< (mv_t)^2 > + < \Delta_\tau^2 > \tag{5.15}$$

The translational kinteic energy growth rate is then given by

$$\frac{dE}{dt} = \frac{< \Delta_\tau^2 >}{2m\tau} \tag{5.16}$$

So the problem remains to evaluate the average of the square of the induced fluctuating impulse Δ_τ. This was done first for thermal radiation by Einstein and Hopf and then adapted to ZPF of SED by Boyer [de Lapena 1995] Thus after some simple calculations we get

$$\begin{aligned}
\Delta_\tau &= \int_0^\tau \sum_{\lambda_1=1}^{2}\sum_{\lambda_2=2}^{2}\int d^3k_1 \int d^3\, k_2\epsilon_z(\bar{k}_1,\lambda_1)\epsilon_z(\bar{k}_2,\lambda_2) \\
& \quad k_1x\frac{3c^3}{2\omega_2^3}.\eta(\omega_1)\eta(\omega_2)\sin\alpha(\omega_2)\cos(\omega_1t - \bar{k}_1.x_1 - \theta(\bar{k}_1,\lambda_1)) \\
& \times \cos(\omega_2t - \bar{k}_2.x - \theta(\bar{k}_2,\lambda_2))
\end{aligned} \tag{5.17}$$

where ω_i is written for $\omega_{\bar{k}_i}$, $i = 1, 2$.

By choosing the origin properly the terms $\bar{k}.x$ can be made small and in particular, for wavelengths of the ZPF much longer than the size of the polarizable particles we can apply the time-honored dipole approximation which omits those terms from the phases. This approximation is justified by the fact that a cut-off form factor is implicit for particles of finite size. After detail calculations we obtain

$$< \Delta_\tau^2 >= \frac{4\pi^4 c^4}{5\omega_0^2}\rho^2(\omega_0)\tau \tag{5.18}$$

Let us now extend this result to the three dimensional case: the polarizable particles can both vibrate and translate along any direction in three dimensional space. Then the impulse can be written as

$$\Delta_\tau = \int_0^\tau \bar{F}dt = \int_0^\tau \{(\vec{p}.\nabla)\vec{E} + \frac{1}{c}\vec{\dot{p}} \times \vec{B}\}dt \tag{5.19}$$

The integrand in RHS is just the Lorentz force for three dimensional dipole moment $\vec{p}$. Vector identity yields

$$\vec{p} \times (\nabla \times \vec{E}) = \nabla(\vec{p}.\vec{E}) - (\vec{p}.\nabla)\vec{E} \tag{5.20}$$

Using Maxwell equations, this becomes

$$(\vec{p}.\nabla)\vec{E} = \nabla(\vec{p}.\vec{E}) + \frac{1}{c}\vec{p} \times \frac{\partial \vec{B}}{\partial t} \tag{5.21}$$

Similar to the previous calculations for one dimensional case,

$$< \Delta_\tau^2 >= \frac{27}{5}\frac{\pi^4 c^4}{\omega_0^2}\Gamma_q \rho^2(\omega_0)\tau \tag{5.22}$$

It is evident from the above calculations that when there is no other random radiation involved like thermal and if the collisions with other particles are excluded, the translational kinetic energy grows as

$$\Omega \equiv \frac{dE}{dt} = \frac{< \Delta_\tau^2 >}{2m\tau} = \frac{9}{z_0}\frac{q^2\hbar^2\omega_0^4}{m\mu c^5} \tag{5.23}$$

where m is the total mass of the polarizable particle and μ is the equivalent mass of the entity that oscillates inside. This formalism can be developed to the polarizable particles in random radiation and in ZPF. This means we have to consider Einstein-Hopf drag force. It is subsequently generalized to monopolar particles immersed in the classical and quantum ZPF.

However, we have mentioned at the beginning of this chapter that problem of acceleration of Cosmic rays (CR) primaries is one of the challanging problems in modern astroparticle physics. So we should deal with concrete particles like protons and nuclei within this framework. We should mention that a cut-off occurs around $\omega = \frac{\pi c}{R}$ where R is the radius of the particle. For an electron $R \simeq 10^{-17}$ cm, the cut-off is around $\omega \approx \frac{\pi c}{R}$. Protons are formed by three quarks which most naturally correspond to the subparticles. In the above framework these subparticles may interact electromagnetically with ZPF. Here, the quarks may be considered as monopolar particles as for electrons and the cut-off $\omega \approx \frac{\pi c}{R_q}$ with R_q as the classical radius of quark.

There should then be many modes of vibration as characteristic of internal oscillations (states) of the proton in which "motion" quarks are involved. It is possible to show [Rueda 1992] that the ultrarelativistic vibrational velocity of these internal motions can not contribute to the translational kinetic energy growth of the particle. They may contribute only to the jiggling internal vibration characteristics of that particular mode. These internal vibrations induced by external oscillating EM fields are substantially present only if the particle has a reasonably large polarizability. Several considerations from quantum chromodynamics (QCD) support the idea that the proton polarizability is not too large. In the classical model that obeys BME, we get the state of polarizability α_p in the zero frequency limit as

$$\alpha_p = \frac{p}{E} = \frac{2\omega_0^2}{3\Gamma_q c^3}; \quad \text{where} \quad \Gamma_q = \frac{2q^2}{3mc^3}$$

Here q should differ much from $e/3$ or $2e/3$ in absolute value. m represents an equivalent mass representing perhaps sizable part of the proton mass, M, say, $M/2$ or $M/3$. For the sake of definiteness we take it as $\frac{M}{3}$. Putting this value in α_p the estimated resonance frequency for the lowest vibrational mode of the protons becomes $\omega_0 \simeq 2.25 \times 10^{23} s^{-1}$. On the otherhand, the radius of the proton is of the order of $r \simeq 10^{-3}$ cm. The corresponding velocities are then $\simeq r\omega_0 = 2.25 \times 10^{10} cms^{-1}$ i.e., in the ultrarelativistic region. Moreover ω_0 is much larger if we associate it with the first excited state mentioned above. Again, nuclei have much larger polarizabilities than protons. Now if we consider that CR acceleration is due to a polarization effect, proton primaries would be relatively absent in comparison with nuclei which is in contrary to conservation. The above approach has been generalized in various directions considering various losses. Another problem with this framework is that it predicts a sharp cut-off for the spectrum of the primary proton at energy $E_0 \geq 10^{19} eV$ in contrary to the experimental data which clearly indicates the absence of any such cut-off Hayashida,1994].

The cut off of the spectrum of primary proton has been shown to be avoided at $E_0 \geq 10^{19}$ eV if one considers the fluctuation of space-time at this ultra-high energy. It raises new interest to study the acceleration of the particles due to the fluctuation of the metric tensor. Again ZPF as well as the irregular magnetic field in Inter Galactic Space (IGS) induces probabilistic aspect in the metric of the background space-time [Roy 1992]. So, a general scheme can be worked out where the ZPF acceleration mechanism, the acceleration produced by irregular magnetic field and the fluctuation of space-time metric can be studied in a unified manner. Any acceptable acceleration mechanism should be capable of accounting for the accelerated particle energy spectra and total numbers, the electron to proton ratios and the acceleration time scales [Ramaty 1987].

In the previous chapter, we have considered the fluctuation in fifth dimension of 4-dimensional Minkowski space as [Sinha 1987]

$$dS^2 = \eta_{uv} dx^u dx^v + \eta_{55} dx^5 dx^5 \tag{5.24}$$

where η_{55} is the random component and $\eta_{\mu\nu} = diag(+1, -1, -1, -1)$.

Physically it means that whenever a particle is considered in this fluctuating vacuum it will be propagated along the direction of this fifth component. It appears that there is a frozen degree of freedom in the vacuum. We have already shown that if this fluctuation is considered along the fifth dimension of a five dimensional space-time then the quantum potential can be generated from this fluctuation. This fluctuation can be shown to have non- classical character which is responsible for the quantum behavior of the microparticles. In that case the metric tensor can be written as

$$g_{\mu\nu} = (1 + \frac{h^2}{2m^2c^2} \frac{\Box\phi}{\phi}) \eta_{\mu\nu} \tag{5.25}$$

where $\mu, \nu = 1, 2, 3, 4$ and quantum potential $= -\frac{h^2}{2m^2c^2} \frac{\Box\phi}{\phi}$. Then it may be considered as conformally flat space-time. The action for a free particle acquires the form

$$S = -mc \int_a^b dS \tag{5.26}$$

where

$$dS = (1 + \frac{h^2}{2m^2c^2}\frac{\Box\phi}{\phi})^{1/2};\ dS_0 \qquad dS_0 = (dx^\mu dx_\mu)^{1/2} \tag{5.27}$$

Following Namsrai [Namsrai 1986], we have the following equation of motion

$$\frac{dU^0_\mu}{dS_0} - \frac{\partial[ln(1+\frac{h^2}{2m^2c^2}\frac{\Box\phi}{\phi})^{1/2}]}{\partial x_\mu} + U^0_\mu(U^\lambda_0 \frac{\partial[ln(1+\frac{h^2}{2m^2c^2}\frac{\Box\phi}{\phi})^{1/2}]}{\partial x_\lambda}) = 0 \tag{5.28}$$

with $U^0_\mu = \frac{dx_\mu}{\partial S_0}$

For simplicity, if we solve the above equation in two dimension then

$$\frac{dV}{dt} = \in (1 - V^2) \tag{5.29}$$

where

$$\in = -C\frac{\partial}{\partial x} ln\left\{1 + \frac{h^2}{2m^2c^2}\frac{1}{\phi}(\frac{\partial^2\phi}{\partial x^2} - \frac{1}{c^2}\frac{\partial^2\phi}{\partial t^2})\right\}^{1/2} \tag{5.30}$$

and $\phi = \phi(x,t)$. Then,

$$\in = \frac{h^2}{4m^2c}\frac{\left\{\frac{1}{\phi^2}\frac{\partial^2\phi}{\partial x^2} - \frac{\partial^2\phi}{c^2\partial t^2}\right\}}{\left\{1 + \frac{h^2}{2m^2c^2}\frac{1}{\phi}(\frac{\partial^2\phi}{\partial x^2} - \frac{\partial^2\phi}{c^2\partial t^2})\right\}}$$

After simple calculations, $V(t)$ can be written as

$$V(t) = \tanh\{\in t + arc\tanh(V_0/C)\} \tag{5.31}$$

or,

$$\dot{V}(t) = \in sech^2\{\in t + arc\tanh(V_0/t)\} \tag{5.32}$$

Now it is evident that the particle in cosmic rays arising from the sources, say, from the Big Bang at the moment at $t = 0$ with initial velocity V_0, accelerates by fluctuation in metric and achieves a very high velocity $V \sim c$ during the evolution of the universe. It is to be noted that in the limit

$h \to 0$, acceleration stops *i.e.*, the particle will move with constant velocity. But at ultrahigh energy, one of the important dissipative mechanisms is the energy loss due to expansion of the universe.

It can be shown that the fluctuation of the metric may be such that the loss of energy will be very less at ultra high energy. In this picture, the cut off of ultra-high energy primary protons can be avoided at $E_0 \geq 10^{19}$ eV in the following manner.

5.2 Highest Energy Cosmic Ray Spectrum

The recent experimental data on the spectrum of primary cosmic rays at energies above 10^{20} eV, as compiled by several groups [Bhattacharjee 1994], clearly indicate the absence of any cut-off at $E_0 \geq 10^{19}$ eV, as suggested by Greisen, Zatsepin and Kuzmin (GZK) [Greisen 1966]. Infact the number of protons of such energy should be very small since an important role is assumed for them by the deceleration due to photoproduction of pions in interactions with the relict radio photons with $T = 3K$. Since, in the laboratory system of the Universe, black-body photons of the universal microwave radiation have an energy of $0.7.10eV$, the Lorentz factor of the cosmic rays needed to boost these photons to the γ-ray energies necessary for interactions to occur are of the order of 10^9 to 6.10^{11} corresponding to cosmic-ray energies in the range of 10^{18}to5.10^{20}eV/nucleon. Moreover, if these protons fill the Universe uniformly, two features are expected to be observed in the spectrum :

(a) a shallow depression at 10^{18}eV caused by e^+e^- production and (b) a pronounced steepening above 5.10^{19}eV from photo meson production. If there are heavy nuclei, i.e. originally iron,a sharp cut-off in the spectrum at 10^{19}eV would be expected, again in conflict with the observation. Several hypothesis [Sinha 1985] have been put forward in order to account for the absence of GZK cut-off. Either they are not nuclei, but something else, i.e. dust particles, high-energy photons, or high energy neutrinos. Or, whether the formulae and concepts based on experiments with much smaller Lorentz factors are applicable to such particles is not cleared till today.

In the first case, the interactions of high-energy neutrinos with nucleons have been studied within Salam-Weinberg model, but the conservative upperlimit gives a cross-section for neutrino interactions at 2.10^{20}eV of $\sigma_\nu \leq 10^{-32}\text{cm}^2$. Soviet school tried to explain it also by generalizing the relativistic laws. They assumed the violation of relativistic relations at sufficiently high Lorentz factor i.e., $\gamma_p = \frac{E_p}{m_pc^2} \approx 10^{11}$ [Kirzhnits,1972 ; Sazanov, 1992]. This kind of frame dependence has been considered by Ingraham in the context of S-matrix and relativistic quantum field theory [Ingraham, 1985].

The main ideas of Soviet school can be summerized as follows: It is possible to generalize the conventional relativistic theory without contradicting the basic postulates of special theory of relativity. Here, the 4-dimensional momentum spcae is assumed to be anisotropic i.e., pseudo-Euclidean space-time is replaced by the Finsler space. We shall consider a 4-vector which does not transform according to the usual vector transformation law when the reference system is changed, but remains constant. Due to this fact, the kinematics is also changed, particularly the relation between cross-sections in different reference systems. In the present approach a generalized framework can be constructed which can accomodate high enough Lorentz factor as well as to satisfy the following general conditions :

(i) The law of inertia : the motion of a free particle in an inertial reference system is rectilinear and uniform

(ii) The principle of relativity : the laws of motion are the same in all inertial frames

(iii) The velocity of light is the same in all reference systems.

(iv) The homogeneity and isotropy of three-dimensional space and time seperately.

(v) The correspondence principle : at relatively low velocities one must again get the conventional theory back.

We can have deviation from the last postulate by postulating the existence of a 4-vector which has the same form in all reference frames. Geometrically, such an anisotropy is in essence equivalent to the transition from the usual pseudo- Euclidean space to Finsler space. In this situation, the coordinate transformation on going to another frame of reference is no longer a point transformation but become a contact transformation. Essentially we get a canonical transformation of general form. However, the transformation of energy-momentum remains a point transformation, although it becomes non-linear. Since the metric of the Finsler space is described by a homogeneous form of the same degree as in the conventional geometry, the difference between the model considered and the existing theory can be determined by a dimensionless parameter.

In the conventional framework, let us consider a Lagangian in 3-dimensional space as

$$L = L(v^2)$$

with v as velocity and momentum as $p_\alpha = \frac{\partial L}{\partial v_\alpha}$. The dispersion relation takes the form

$$I(p_i) = m^2$$

where $I(p_i)$ is a homogeneous function of second degree in p_i. Let us restrict ourselves to functions of the form $I(p_i)$, i.e.,

$$I(p_i) = f(p_i)(E^2 - \vec{p}^2) \tag{5.33}$$

where $f(p_i)$ is a homogeneous positive function of the momenta of zero degree. For small velocities $f(p_i) = 1$ as in the usual theory. If we consider the isotropy of three-dimensional momentum space, the only argument upon which $f(p_i)$ may depend is $\frac{p^2}{E^2}$. Let us now derive the expression for the velocity

$$\begin{aligned} v^\alpha &= \frac{\partial E}{\partial p_\alpha} \\ v &= u\Psi(u^2) \\ \Psi(u^2) &= \frac{1 - (1 - u^2)'}{(ln f)} \end{aligned} \tag{5.34}$$

Here, $u = \frac{p}{E}$ and the prime stands for derivative with respect to u^2. Then we can write energy and momentum as

$$\begin{aligned} E &= \frac{m}{\sqrt{(1 - n^2)f(n^2)}} \\ p &= \frac{mu}{\sqrt{(1 - n^2)f(n^2)}} \end{aligned} \tag{5.35}$$

respectively. In the limit $u \to 0$ and $v \to 0$, we get the conventional expression.

Now the law of transformation of energy and momentum from one frame of reference to other can be directly obtained from the invariance condition of the dispersion law . But the main problem is how to determine the form of the function, f ?

Analyzing the experimental situations, we can show that f differs from unity not more than 10^{-4} if $1 - \frac{p^2}{E^2} \geq 10^{-8}$. Furthermore, if we want to attribute the absence of a break in the spectrum of CR at an energy $E_k \sim 5\times10^{19}$ to the deviation of f from unity, we have to assume that the difference $(f-1)$ becomes equal to unity at this energy. This corresponds to $\lambda = 1 - (\frac{p^2}{c^2}) \sim 10^{-22}$. Hence, the function is equal to unity with high precission in most of its range and changes abruptly at the edge of this domain. This form of f reflects a slow convergence of the expansion in the region of interest. In this situation we can think over another dimensionless argument

$$\xi = \frac{p^2}{(E^2 - p^2)}$$

in terms of which the function behaves much more smoothly. Here, the convergence of the expansion of $f(\xi)$ is faster,

$$f(\xi) = 1 + c_1\lambda\xi + c_2\lambda^2\xi^2 + \cdots$$

where c_1 and c_2 are numerical coefficients of the order of unity. Considering the above expression for small momenta, it follows that we have to omit the linear term. By putting $c_2\,\lambda^2 \equiv \alpha$, we obtain

$$f = 1 + \alpha\xi^2 + \cdots \quad \text{or} \quad f = 1 + \alpha\gamma^4 + \cdots \tag{5.36}$$

taking $|\,\alpha\,| \sim 10^{-44}$ and $\alpha\gamma^4 < 1$.

We assume that the function f has a finite limit $f(\alpha)$ as $\xi \to \infty$. For a photon in particular, $f = f(\infty)$ and its dispersion law does not differ from the conventional one. Let us consider the influence of this function f on the ultra-high energy proton spectrum. In the conventional theory the cut-off of CR spectrum due to intense photoproduction of pions starts at proton energies for which the background photon has in the rest system of the proton an energy of the order of the pion mass. It is important to calculate the proton lifetime with respect to photoproduction. This can be computed if we can calculate correctly the statistical factor represented by the Planck distribution of background photons

$$H = \exp[-\frac{\omega_c}{2\gamma_p KT}] \tag{5.37}$$

where ω is the energy of photons in proton rest system. This factor, written in the proton rest system, takes the form

$$H = \exp(-\frac{\omega_c}{2\gamma_p KT}) \tag{5.38}$$

for a central collision.

γ_p is the Lorentz factor of the proton and ω_c is the photon energy in this system ($\omega_c \geq m_\pi$). For $\gamma_p \geq \frac{\omega_c}{2\pi KT}$, this factor rises sharply, leading to a rapid decrease in the proton lifetime. Now using our generalized transformation for ω_c and ω, we get

$$\omega_c = \sqrt{f(\xi_p)2\gamma_p\omega} \tag{5.39}$$

Consequently,

$$H = \exp\left(-\frac{\omega_c}{2\gamma_p KT\sqrt{f(\xi_p)}}\right) \tag{5.40}$$

If α is negative in the expression of $f(\xi_p)$, the presence of supplementary factor in the argument of the exponential function leads to a decrease in the effective temperature of photons and hence to decrease in the proton lifetime. In order to eliminate completely the cut-off in the CR spectrum at arbitrary values of γ, it is sufficient to take $f(\infty) \sim 0.1 - 0.01$ which is quite possible within this framework.

The frame dependence of S-matrix operator within the framework of relativistic quantum field theory by Ingraham [Ingraham 1985] helps us to understand the absence of GZK cut-off in a more rigorous way. It has been found that the physical origin of this frame dependence can be treated back to its origin to random space-time points. Recent results of COBE indicates the presence of fluctuations in Cosmic Microwave Background Radiation (CMBR) just above the limit where the conventional theories would have run into trouble. This temperature fluctuations are taken as surrogate for density perturbations in the early universe which in turn perturb the metric of the space-time and thus the frame dependence can be envisaged. It is to be noted that random background electromagnetic field may induce [Roy 1992] a probabilistic nature in the background space-time metric.

Let $(\frac{d\sigma}{d\Omega})_{L'}$ and $(\frac{d\sigma}{d\Omega})_L$ denote the differential cross-section for a particular interaction as actually measured in the frames L' and L where, L' and L are laboratory and c.m. system respectively. These cross-sections can be constructed out of the S-matrix elements $S(L')$ and $S(L)$. With $S(K') \neq S(K)$, now the amplitudes for the same scattering experiment in the frames L and L' will be

$$<\phi|S(L)|K> \quad \text{and} \quad <\phi|S(L')|K'>$$

respectively. The cross-sections as measured in the laboratory frame L' and transformed to c.m. frame L are

$$\left|\frac{d\sigma}{d\Omega}\right|_L = \left|\frac{d\sigma}{d\Omega'}\right|_{L'} \frac{d\Omega'}{d\Omega} \tag{5.41}$$

where, $\frac{d\Omega'}{d\Omega}$ is the lab-cm. Jacobian. But we know

$$\left|\frac{d\sigma}{d\Omega}\right|_{L'} = \left|\frac{d\sigma}{d\Omega}\right|_L \tag{5.42}$$

is satisfied in the usual scheme of particle physics because S is taken as frame independent. However, it is not satisfied in the generalized scheme as

$$q_1^2(L') \neq q_1^2(L) \tag{5.43}$$

for a given four momentum q and $q_1^2(L)$ being the square of the spatial part of the four vector q in the frame L, is

$$q_1^2 = q^2 + [\vec{q}.\vec{n}(L)]^2 \tag{5.44}$$

in which $\vec{n}(L)$ can be thought of as unit time-like vector directed along L's positive time axis i.e.,

$$\begin{aligned} n^2(L) &= -1, \quad n^0(L) > 0 \\ n^\mu(L) &= (1,0,0,0) \end{aligned} \tag{5.45}$$

referred to frame L.

Due to the above considerations, relativistic kinematic relations can be written as

$$\frac{E^{*^2}}{c^2} - p^{*^2} = (1+\alpha^2)\left[\frac{E^2}{c^2} - p^2\right] \tag{5.46}$$

with (E^*, p^*) and (E, p) being in lab system and c.m. system respectively and α, a parameter related to $n(L)$.

It is to be mentioned that the dominant primary component of the cosmic ray at $E \geq 10^{19}$ eV is considered to be as protons. But the number of protons at such energy in cosmic rays should be very small since photo-production of pions starts at such high energy in the interaction of protons with the relict radiophotons with $T = 2.7^o K$. In the conventional theory, the cut-off of the cosmic ray spectrum due to intense photoproduction of pions starts at proton energies for which this background proton has in the rest system of the proton, an energy of the order of the pion mass (in case of central collisions). So, it is important to calculate the proton life time with respect to photoproduction. The important factor in calculating proton life time is the statistical factor represented by Planck distribution of background protons i.e.,

$$H = \exp(-\frac{\hbar\omega_e}{KT})$$

where $\hbar\omega_e$ is the energy of the protons in the earth reference frame and the proton distribution is measured (henceforth call it as e-system). This factor H, when written in the proton rest system (call it l-system), takes the form

$$H = \exp(-\frac{\hbar\omega_l}{2\gamma_p KT}) \tag{5.47}$$

for a central collision. Here, γ_p is the Lorentz factor of the proton and ω_l is the proton energy in the l-system ($\hbar\omega_l \geq m_\pi c^2$). The Lorentz factor γ_p which connects the "earth" reference frame or e-system and proton rest system or l-system is

$$\gamma_p \simeq \frac{E_p}{m_{pc^2}} \sim 10^{11} \tag{5.48}$$

for $E_p \geq 10^{19} eV$. As γ_p is very large with respect to that measured for highest energy available so far in Laboratory experiment, Soviet school expressed their doubt regarding the validity of relativistic kinematics at such ultra high energy.

Now in our frame dependence S-matrix formalism,

$$\left|\frac{d\sigma}{d\Omega}\right|_{l'} \neq \left|\frac{d\sigma}{d\Omega}\right|_{l} \tag{5.49}$$

for $\gamma + p$ interaction and the cross-section in l-system will be undefined accompanied by a Gaussian form factor involving $q_1^2(L)l^2$ where l denotes the length scale of fluctuation of $CMBR$.

Taking

$$\begin{aligned} \left|\frac{\Delta T}{T}\right| &= \left|\alpha^2\right| \simeq \left|\frac{\Delta\rho}{\rho}\right| \\ &\simeq 10^{-5} \end{aligned} \tag{5.50}$$

as observed from *COBE* measurements, the scale of fluctuation l can be estimated from

$$\alpha = \frac{l_p^2}{l^2} \tag{5.51}$$

as of the order of 10^{-30}cm, taking l_p as Planck length. This is consistent with the value already estimated by Blokhintsev from different combination. For microwave radiation, $\omega_e \sim 10^{10}$rads^{-1}. So,

$$w_l \sim 2\gamma_p w_e = 10^{21}\text{rads}^{-1}$$

Now the Planckian distribution in l-system will be modified as [Kirzhnits, 1972]

$$H = \exp[-\frac{\hbar w_l}{2\gamma_p KT}(1+\alpha^2)] \tag{5.52}$$

Here, the second term in H suppresses the interaction losses for protons and hence counteracts the spectrum cut-off. It explains the absence of cut-off in the primary cosmic ray spectrum at the highest energy available till now. It is interesting to note that if we consider the interactions of high energy neutrinos with nucleons in the framework of stochastic space-time, then the corss-section will not be negligible in contrast to Salam-Weinberg theory. The neutrinos may arise from the decay of the charged pions produced in interaction such as

$$\begin{cases} p + \pi_0 \\ n + \pi^+ \end{cases}$$

They have energies in the observer frame of refernce which are typically one order of magnitude smaller than that of original protons. At neutrino energy up to 10^{11} eV, the total cross-section for neutrino-nucleon interaction of the charged-current type

$$\nu + \mathcal{N} \rightarrow +\text{hadrons}$$

is linear in incident neutrino energy. If this linear rise in cross-section continues up to 10^{20}eV, the cross-section would be of the order of a millibarn. Neutrino interactions in the atmosphere could then produce air showers. The linear rise in the cross-section with energy holds only in the case of the four-fermion coupling. Now in the stachastic space-time we can write the metric as

$$G_{\mu\nu} = \delta_{\mu\nu} + \alpha g_{\mu\nu}$$

where α is the parameter indicating the deviation from Euclidean structure as described in Chapter **3**. Then the neutrino cross-section can be written as

$$\sigma_\nu \simeq \beta_\nu E_\nu + \gamma^2 E_\nu$$

with

$$\beta_\nu \simeq 0.8.10^{-38}\text{cm}^2(\text{GeV})^{-1}$$

Here $\gamma^2 \simeq \alpha^2/M_p^2 \equiv 10^{-6}\text{cm}^2(\text{GeV})^{-1}$.

So at ultrahigh energy, Ist term can be neglected and only the 2nd term dominates. Now if we normalise $\frac{\alpha^2}{M_p^2}$ with M_p being the proton mass, then

$$\sigma_\nu \lesssim 10^4\text{cm}^2$$

It thus appears that cut-off can be avoided in this framework which was not possible in the unified gauge theory of Salam and Weinberg. Ofcourse we have not considered yet the status of galactic and supercluster hypothesis about the origin in the shape of energy spectrum. However, it is clear from the above analysis that there will be no cut-off at such ultra high energy as predicted by GZ

5.3 Propagation of light in a dissipative Maxwell Vacuum :

Several authors studied Maxwell equations [Vigier 1990; Kar 1993] in vacuum with non-zero conductivity coefficient i.e., with $\sigma \neq 0$. The non-zero conductivity coefficient gives rise to a dissipative term in the field equation. Now due to the presence of the dissipative term in the field equation, the photon looses energy during the propagation through this vacuum. This dissipation can be related to the fluctuation of the refractive index of the underlying vacuum. In this model of fluctuating vacuum, the velocity of the propagation of the disturbance is shown to be finite. This has great significance in the microscopic domain of space-time. **(a) Maxwells Equation in Vacuo with $\sigma \neq 0$:**

If we endow the vacuum with non-zero conductivity coefficient $\sigma \neq 0$, Maxwell equations can be written in the form

$$\left.\begin{aligned} \nabla.\vec{E} &= 0 \\ \nabla \times \vec{H} &= \sigma\vec{E} + \epsilon_0\chi_e\frac{\delta\vec{E}}{\delta t} \\ \nabla.\vec{H} &= 0 \\ \nabla \times \vec{E} &= -\mu_0\chi_m\frac{\partial\vec{H}}{\partial t} \end{aligned}\right\} \qquad (5.53)$$

where ϵ_0 denotes vacuum's dielectric constant
μ_0 denotes vacuum's permeability constant
χ_e is the relative dielectric constant
χ_m is the relative permeability constant.

In the usual classical vacuum, one generally has $\sigma = 0$; $\nabla\times\nabla\times\vec{E} = -\nabla^2\vec{E}$. So,

$$\nabla^2\vec{E} = -\frac{\epsilon_0\chi_e\chi_m}{c^2}\frac{\partial^2\vec{E}}{\partial t^2} + \sigma\mu_0\chi_m\frac{\partial\vec{E}}{\partial t}. \tag{5.54}$$

If we consider plane waves in the z-direction i.e.,

$$E_x = b\exp[iw(t - \frac{z}{v})] \tag{5.55}$$

$$H_y = b(\frac{\epsilon_0\chi_e}{\mu_0\chi_m})\exp\left[iw(t - \frac{z}{v})\right] \tag{5.56}$$

putting $q = \frac{1}{v}$ in the plane wave solutions of E, we get

$$q^2 = \frac{\chi_e\chi_m}{c^2}\left(1 - \frac{i\sigma}{\epsilon_0\chi_e\omega}\right)\epsilon_0\mu_0 \tag{5.57}$$

Here, q can be considered as complex in nature having the form $\alpha - i\beta$ where, α & β are real and given by

$$\alpha = 1 + \frac{1}{8}(\frac{\sigma^2}{\epsilon_0^2\chi_e^2}).\frac{1}{w^2} + O(\frac{\sigma^4}{w^4}) \tag{5.58}$$

and

$$\beta^2 = \frac{1}{2}(\frac{\sigma}{\epsilon_0\chi_e})^2.\frac{1}{w^2} \tag{5.59}$$

for $(\frac{\sigma}{w}) \to 0$. The velocity defined by v will give rise to complex refractive index η in the vacuum. The velocity defined $V = \frac{1}{\alpha}$ is the phase velocity of propagation of of disturbance through the underlying vacuum. Henceforth, it will be denoted as V_p. After simple calculation, the phase velocity can be written as

$$\begin{aligned} V_p &= \frac{1}{\alpha} \\ &= \frac{c}{(\chi_e\chi_m)^{\frac{1}{2}}}\left[1 - \frac{1}{8}\frac{\sigma^2}{\epsilon_0\chi_e^2 w^2}\right] \end{aligned} \tag{5.60}$$

and the group velocity

$$V_g = \frac{c}{(\chi_e \chi_m)^{\frac{1}{2}}} \left[1 + \frac{1}{8} \frac{\sigma^2}{(\epsilon_0 \chi_e)^2 w^2} \right] \tag{5.61}$$

In this case, the phase velocity is less than the speed of light but the group velocity may be greater than c. We will use this phase velocity in calculating non-zero mass of photon, since according to Sommerfeld and Brillouin [Brillouin 1960], it is the front velocity (not group velocity) to be relevant for explaining Einstein's causality condition. Moreover, Maxwell theory predicts that electromagnetic waves in vacuum have always a constant front velocity equal to c.

Again, taking the above calculated values of α and β in E_x and H_y, the two following cases may arise :

(i) Plane waves are progressively damped with the decay factor $\exp(-Kz)$ where $K = w\beta$.

(ii) The velocity of propagation of the wave is given by V_g and it varies with the frequency.

Using de Broglie relation

$$E = \hbar\nu = \frac{m_\gamma c^2}{(1 - \frac{V_p^2}{c^2})^{\frac{1}{2}}} \tag{5.62}$$

m_γ, the non-zero photon mass, then becomes

$$m_\gamma = \frac{\sigma h}{2\sqrt{2}\pi(\epsilon_0 \chi_e)} = \frac{\hbar H}{2c^2} \simeq 10^{-65}\text{gm} \tag{5.63}$$

where H denotes Hubble's constant [Fuli 1981].

Fuli [Fuli,1981] made an estimate for non-zero rest mass of photon in the following manner. Here, the interaction of the photon with gravitational field is considered. This interaction causes to loose energy in the form of very low energy secondary photons and the Heisenberg uncertainty principle can be applied to one of the secondary photon to estimate the rate of energy loss by the primary photon. Following this argumernt one has

$$\frac{dh\nu}{dt} = -\frac{h\nu}{t_c}; \quad h\nu = h_0\nu_0[-\frac{r}{r_c}] \tag{5.64}$$

where h_0 and ν_0 are the Planck constant and the photon frequency respectively at $t = 0$ and $t_c = \frac{h}{2m_0c^2}, r_c = \frac{h}{2m_0c}, r = t_c, m_0$ is the photon rest mass and c is the speed of light. We assume $h_0 = h$. So the red shift relation is

$$z = \frac{\nu_0 - \nu}{\nu} = \exp[\frac{r}{r_c}] - 1 \qquad (5.65)$$

r_c is the range of electromagnetic interaction when

$$r \ll r_c; \quad z \approx \frac{2m_0c}{h}r$$

It is also assumed that Hubble red shift relation is only justified for

$$r \ll r_c, \quad \text{i.e,} \quad z = \frac{H}{c}r$$

with H as Hubble constant.

Comparing the above two equation, we get

$$m_0 = \frac{hH}{2c^2} \qquad (5.66)$$

Now substituting the value of h, H and c^2, Fuli got the photon rest mass as

$$\approx 10^{-66} gm$$

In the limit $h \to 0$; $m_0 \simeq 0$. Again in the nonrelativistic limit, $c \to \infty$ and $m_0 \simeq 0$. It should be noted that the ratio of the electron rest mass to the photon rest mass showed here to be of $\sim 10^{39}$. This is just a Dirac large number.

From the above relations, we get an exponential red shift relation as

$$z = \exp[\frac{H}{c}r] - 1 \qquad (5.67)$$

The distance between the Earth and a galaxy with $z \approx 3.53$ is $1.5.10^{10}$ light years. The range of the electromagnetic interaction is

$$r_c = \frac{2c}{H} \simeq 2.10^{10}$$

light years. which value has been called the radius of the universe. Thus it can be proposed that photon has an effective rest mass in vacuum defined by the coefficient $\sigma \neq 0$, which one can consider as being constant except in the neighborhood of strong gravitational sources which might increase the vacuum's density. It is now clear from the above analysis that the progressive waves will be heavily damped for complex refractive index of the vacuum and we get the non-zero rest mass of the photon. This dissipation can be related to the fluctuation of the medium.

(b) Waves in the Einstein-de Broglie's Theory of Light :

In ordinary conditions, one is brought back to de Broglie theory of light [De Brogli 1972] which gives (in vacuum) with $k_0 = \frac{m_\gamma c}{\hbar}$, the relations :

$$\begin{cases} \vec{E} & = -\frac{1}{c}\frac{\partial \vec{A}}{\partial t} - \text{grad } V \\ -\frac{1}{c}\frac{\partial \vec{H}}{\partial t} & = \nabla \times \vec{E} \\ \frac{1}{c}\frac{\partial \vec{E}}{\partial t} & = \nabla \times \vec{H} + k_0^2 \vec{A} \end{cases} \quad \left\{ \begin{array}{lcl} \vec{H} & = & \nabla \times \vec{A} \\ \nabla . \vec{H} & = & 0 \\ \nabla . \vec{E} & = & -k_0^2 V \end{array} \right\} \tag{5.68}$$

$$-\frac{1}{c}\frac{\partial V}{\partial t} + \nabla . \vec{A} = 0.$$

It follows immediately that for quantities H, E, A and V, one has in vacuum the set of equations

$$F_{ik} = \frac{\partial A_k}{\partial x_i} - \frac{\partial A_i}{\partial x_k}; \quad \partial^k F_{ki} = -k_0^2 Ak \; ; \quad \frac{\partial A_i}{\theta x_i} = 0 \tag{5.69}$$

and

$$\frac{\partial F_{ik}}{\partial x_l} + \frac{\partial F_{kl}}{\partial x_i} + \frac{\partial F_{li}}{\partial x_k} = 0 \tag{5.70}$$

which imply, with F denoting any of the ten quantities $A_x, \ldots, E_z$, the relations

$$\Box F + k_0^2 F = 0 \tag{5.71}$$

Writing, $P = e^{i(k_0 ct - \vec{k}.\vec{r})}$, we obtain for the correspondence between the mechanical quantities (energy W and impulsion P) of the photons, the relation $K = (\frac{2\pi}{\hbar})\omega$ which satisfy the well known relativistic equation

$$W^2/c^2 = |P|^2 + {m_\gamma}^2 c^2$$

is valid for the energy (W) and impulsion (P) of individual photons.

The general plane-wave solutions has been explicitly calculated by de Broglie. For a plane wave moving in the z-direction,

$$A_x = \frac{C_1 + C_2}{2}P; \quad A_y = \frac{C_1 - C_2}{2}iP; \quad A_z \simeq C_3P; \quad V = \frac{C_3|k|}{k}P.$$

$$E_x \simeq -ik\frac{C_1 + C_2}{2}P; \quad E_y = k\frac{q - C_2}{2}P \quad E_z = -i\frac{k_0^2}{k}C_3P$$

$$H_x = |k|\frac{C_1 - C_2}{2}P; \quad H_y = -i|k|\frac{C_1 + C_2}{2}P; \quad H_z = 0$$

C_i being arbitrary constants.

For plane T (transverse waves), ($J_3 = \pm 1$),

$$E^T = -i(\frac{P}{m_\gamma})A^T \quad \text{and} \quad |H^T|/|E^T| = k_0/\nu \tag{5.72}$$

are orthogonal between them, with $A^T || E^T$ and $V^T = 0$. The density of the wave's energy is

$$W^T = |E^T|^2 + |H^T|^2 - (|A^T|^2 + (|V_T|^2) \tag{5.73}$$

and A^T, $V^T << E^T$, H^T. The solutions can be split into four independent monochromatic waves i.e.,

- (i) A Maxwellian transverse right-rotated wave D
- (ii) A Maxwellian transverse left-rotated wave G
- (iii) A Maxwellian longitudinal wave L
- (iv) A non-Maxwellian wave.

One can also normalize such waves in a finite volume V, which give ($k > 0$)

$$|C_1| = (\hbar c/2\pi kV)^{\frac{1}{2}}, \quad |C_3| = (\hbar ck/4\pi k_0^2V)^{\frac{1}{2}}$$

$$|C_2| = (\hbar/2\pi kV)^{\frac{1}{2}}, \quad |C_4| = (2k_0/kV)^{\frac{1}{2}}.$$

and define the linearly polarized waves.

(c) Extended Electromagnetic Theory :

Recently, Bo Lehnert [Lehnert 1995] proposed an extended Lorentz-invariant form of Maxwell equation. His theory is based on the hypothesis that the electric charge density can be interpreted as an intrinsic property of electromagnetic field in vacuo. The basic equations of this extended form are consistent with and related to the Dirac theory of electron. In this framework, Maxwell equation can be written as (in SI units)

$$\left.\begin{aligned}
\nabla \times \frac{\vec{B}}{\mu_0} &= \vec{j} + \epsilon_0 \frac{\partial \vec{E}}{\partial t} \\
\nabla \times \vec{E} &= -\frac{\partial \vec{B}}{\partial t} \\
\vec{j} &= \bar{\rho}\vec{C}, \\
\nabla . \vec{E} &= \bar{\rho}/\epsilon_0
\end{aligned}\right\} \tag{5.74}$$

The additional feature is the possibility for the divergence of the electric field to differ from zero also in vacuo. The form $\vec{j} = \bar{\rho}\vec{C}$ represents an electric polarization current in vacua. The combination of the above equations can be written in the form

$$\left(\frac{\partial^2}{\partial t^2} - c^2\nabla^2\right)\vec{E} + (c^2\nabla + \vec{C}\frac{\partial}{\partial t})(\nabla.\vec{E}) = 0 \tag{5.75}$$

Here, the potentials $\vec{A}$ and ϕ are derived from the sources $\vec{j}$ and ρ through the above equation (5.74), which yields

$$\Box^2(\vec{A}, \frac{i\phi}{c}) = -m_\gamma(\vec{j}, ic\bar{\rho}) \tag{5.76}$$

where $\Box = (\frac{\partial}{\partial x}, \frac{\partial}{\partial y}, \frac{\partial}{\partial z}, \frac{\partial}{ic\partial t})$.

When considered with Lorentz condition, this gives

$$\nabla.\vec{A} + \frac{1}{c^2}\frac{\partial \phi}{\partial t} = 0 \tag{5.77}$$

It should be observed that equation (5.76) is a Proca type equation. Again, Lorentz invariance leads to

$$\vec{j}^2 - c^2\bar{\rho}^2 = \text{constant} = 0; \quad \text{where} \quad \vec{C}^2 = c^2 \tag{5.78}$$

The following limiting cases now may arise :

- (i) When $\nabla.\vec{E} = 0$ and $\nabla \times \vec{E} \neq 0$, the result is a conventional transverse electromagnetic wave (EM).
- (ii) When $\nabla\vec{E} \neq 0$ and $\nabla \times \vec{E} = 0$, a longitudinal purely electric space charge wave arises (S).
- (iii) Where both $\nabla.\vec{E} \neq 0$ and $\nabla \times \vec{E} \neq 0$, a hybrid non-transverse electromagnetic space-charge wave appears (EMS).

The S wave can be considered as a special degenerate form of EMS wave. It can be shown that both EM and S waves are seperate modes which can not be derived from the EMS mode by simple asymptotic processes. A general form of electromagnetic field can be obtained from a superposition of various EM, S and EMS modes. It should be observed that EMS modes can have different field vectors $\vec{C}$ [Lehnert, 1997].

It should be noted that in the approach of Maxwell vacuum with non-zero conductivity we have considered

$$\nabla.\vec{E} = 0 \text{and} \nabla.\vec{E} \neq 0$$

So this is nothing but the limiting case (i) of the above approach as proposed by Lehnert. But in the former approach the photon acquires a non-zero mass as it propagates through Maxwell vacuum with non-zero conductivity. In this case one should associate a longitudinal mode to this kind of photon. On the other hand in Lehnert approach, we get only transverse wave i.e. EM wave under this condition (i). It seems that the two theories are not in agreement with each other. But if we formilate the EBP theory with $\sigma \neq 0$ in a fully relativistic way as well as to make it gauge invariant way, it is necessary to introduce the concept of space-charge in vacuo as given by

$$j = (\vec{j}, j_0); \quad j_0 \neq 0$$

Here j_0 is related to $\bar{\rho}$. In our previous formulation we have taken $j_0 = 0$. Now we have both $\nabla.\vec{E} \neq 0$ and $\nabla.\vec{E} \neq 0$. This is nothing but the case (iii) of Lehnert approach i.e. EMS wave.

(d) Axisymmetric wave modes, Angular momentum and Non-zero photon mass:

We now study axisymmetric wave modes varying as $\exp[i(-\omega t + kz)]$ in a cylindrical frame (r, ϕ, z) of reference where ϕ is an ignorable coordinate. In this frame the velocity vector is assumed to have the form

$$\vec{C} = c(0, \cos\alpha, \sin\alpha) \tag{5.79}$$

with a cosntant α. We further define the operators

$$D_1 = \frac{\partial^2}{\partial r^2} + \frac{1}{r}\frac{\partial}{\partial r} + \frac{\partial^2}{\partial z^2} - \frac{1}{c^2}\frac{\partial^2}{\partial t^2} \tag{5.80}$$

$$D_2 = \frac{\partial}{\partial t} + c(\sin\alpha)\frac{\partial}{\partial z} \tag{5.81}$$

$$D_3 = \frac{\partial^2}{\partial z^2} - \frac{1}{c^2}\frac{\partial^2}{\partial t^2} \tag{5.82}$$

The basic equations then reduce to

$$D_2(\nabla.\vec{E}) = 0 \tag{5.83}$$

$$D_2(D_1 - \frac{1}{r^2})E_r = 0 \tag{5.84}$$

$$D_3(D_1 - \frac{1}{r^2})E_\phi = \frac{1}{c}\cos\alpha\frac{\partial^2 D_1 E_z}{\partial z\partial t}D_1 E_z \tag{5.85}$$

This set of equations corresponds to two branches of solutions:

1. When $D_2 E_r$ and $D_2 E_z$ are different from zero, this represents a classical electromagnetic (EM) mode with vanishing electric field divergence.

2. When $D_2 E_r = D_2 E_z = 0$, this branch represents an electromagnetic space-charge(EMS) mode with non-zero electric field divergence in vacuo. The first branch has no limited solution for entire space.

For axisymmetric EMS modes of the second branch we have $\nabla.\vec{E} \neq 0$, the dispersion relation becomes

$$\omega = kc\sin\alpha \tag{5.86}$$

We now introduce the function

$$G_0.G \equiv E_z + (\cot\alpha)E_\phi; \quad G \equiv R(\rho)\exp[i(-i\omega t + kz)] \tag{5.87}$$

Where G_0 is an amplitude factor and $R(\rho)$ is dimensionless. The operator

$$D = \frac{\partial^2}{\partial \rho^2} + \frac{1}{\rho}\frac{\partial}{\partial \rho} - \xi^2(\cos\alpha)^2 \quad \xi = kr_0 \tag{5.88}$$

is further defined. Then the electric field components become

$$\begin{aligned}
E_r &= -iG_0\xi\frac{\partial}{\partial\rho}[(1-\rho^2D] \\
E_\phi &= G_0(tg\alpha)\rho^2DG \\
E_z &= G_0(1-\rho^2D)G
\end{aligned} \tag{5.89}$$

Similarly the magnetic field components can be written as

$$\begin{aligned}
B_r &= -\left[\frac{1}{c\sin\alpha}\right]E_\phi \\
B_\phi &= \left[\frac{1}{c\sin\alpha}\right]E_r + i\left[\frac{1}{\xi c\sin\alpha}\right]\frac{\partial E_z}{\partial\rho} \\
B_z &= \left[\frac{1}{\xi c\sin\alpha}\right](\frac{\partial}{\partial\rho}+\frac{1}{\rho})E_\phi
\end{aligned} \tag{5.90}$$

Consequently, the function G can be considered as a generating function from which the entire electromagnetic field of an elementary axisymmetric EMS mode can be determined. It is possible to superimpose elementary EMS modes of different wave numbers k in the z direction, to form a travelling wave packet by means of Fourier analysis, and for such a packet to have extensions also in the positive and negative z-directions. Here this is assumed to be the case, and $\vec{E}$ and $\vec{B}$ to be the resulting electric and magnetic fields which then become finite at $r = 0$ and zero at $r \to \infty$ and $z = \pm\infty$. For a packet of this kind to represent a limited spectral line width and a well defined freqeuncy $\omega = kc(\sin\alpha)$, the spatial extension of the packet in the z-direction then has to be much larger than its wave length $\frac{2\pi}{k}$. With these assumptions we first turn to the total integrated charge. With dV and dS as volume and surface elements, it becomes

$$q_0 = \epsilon\int_0^\infty(\nabla.\vec{E})dV = \epsilon_0\int_0^\infty \vec{n}.\vec{E}dS = 0 \tag{5.91}$$

when the electric field vanishes at large distances from the packet. We next consider the integrated magnetic moment which becomes

$$M_0 = \frac{1}{2}\epsilon_0 c(\cos\alpha)\int r(\nabla.\vec{E})dV \tag{5.92}$$

and does not generally disappear. Turning further to the energy density of the electromagnetic field

$$\omega = \frac{1}{2}(\bar{\rho}\phi + \vec{j}.\vec{A}) = \frac{1}{2}\bar{\rho}(\phi + \vec{C}.\vec{A}) \tag{5.93}$$

where $\vec{A}$ and ϕ are the magnetic and electric potentials, the equivalent mass in the frame which allows the propagation of the wave packet becomes

$$m_\gamma = \frac{2\pi}{c^2}\int_0^\infty \int_{-\infty}^{+\infty} r\omega dr dz \tag{5.94}$$

Likewise the integrated angular momentum becomes

$$s_0 = \left[2\pi\frac{\cos\alpha}{c}\right]\int_0^\infty \int_{-\infty}^{+\infty} r^2\omega dr dz \tag{5.95}$$

For a wave packet with finite characteristic dimensions r_0 and z_0 in the r and z directions, and with an energy density ω being proportional to $\bar{G}_0^2$ as obtained from the fields $(\vec{E}, \vec{B})$ and the generating function G, we introduce the normalized quantities $\rho = \frac{r}{r_0}$, $\zeta = \frac{z}{z_0}$ and $\omega(r,z) = \bar{G}_0^2 W(\rho,\zeta)$. Then

$$\begin{aligned}
m_\gamma &= (\frac{2\pi}{c^2}) r_0^2 z_0 \bar{G}_0^2 J_m \\
J_m &= \int_0^\infty \int_{-\infty}^{\infty} \rho W d\rho d\zeta \\
s_0 &= \left[2\pi(\frac{\cos\alpha}{c})\right] r_0^3 z_0 \bar{G}_0^2 J_s \\
J_s &= \int_0^\infty \int_{-\infty}^{+\infty} \rho^2 W d\rho d\zeta
\end{aligned} \tag{5.96}$$

where $\bar{G}_0$ thus stands for an equivalent amplitude factor. Assuming the present system to behave as a boson, we have

$$s_0 = \frac{h}{2\pi}$$

with h denoting the Planck constant and

$$m_\gamma r_0 = \frac{hJ_m}{2\pi c(\cos\alpha)} J_s$$

When requiring m_0 and r_0 to become small. forms of W have to be found which make the ratio J_m/J_s small. That such functions are likely to exist is illustrated by the simple form $\rho^\gamma \exp(-\rho)$. Now let us consider the energy content of a wave packet of narrow line width as well as the asymptotic limit of a monochromatic wave of infinite axial extension and a discrete frequency ω. With the phase and group velocities $v = c(\sin\alpha)$, the energy realtions due to Planck and Einstein then yield

$$\frac{h\omega}{2\pi} = m_\gamma c^2 \left[1 - \frac{v^2}{c^2}\right]^{-\frac{1}{2}} \tag{5.97}$$

This leads to

$$\omega = \frac{J_m c}{J_s r_0 (\cos\alpha)^2} \tag{5.98}$$

With $\cos\alpha \ll 1$ and $\frac{J_m}{J_s} \ll 1$, the frequency ω thus becomes relevant to the present concept of a propagating wave packet with a defined frequency and a small radius r_0. It is to be noted that the above conditions imply that m_γ and $m_\gamma r_0$ become small and at the same time the condition of a quantized angular momentum is being satisfied. Again it should be observed that the introduction of a nonzero but small conductivity in vacuo as described before introduces a small but finite rest mass of the photon. The phase and group velocities have been calculated in this framework. In this approach, the phase velocity is also less than the speed of light. Thus the approach of a non-zero electric field divergence in vacuo can become reconciliable with such a concept if we coose the angle α very close to $\frac{\pi}{2}$ i.e., where $0 < \cos\alpha \ll 1$. Then the phase velocity $c\sin\alpha$ also becomes slightly smaller than c.

(d) Dissipative Maxwell Vacuum :

The vacuum dissipation mechanism associated with non-zero photon mass may be due to the following reasons :

(i) The statistical behavior of microparticles has been linked to the fluctuation of space-time metric. This fluctuation is said to be universal and real statistical fluctuations. It reflects the basic stochastic character of Dirac's aether. However, since they result from a process which is superimposed on the evolution of the local dynamical variables and introduce an element of randomness into the evolution of the dynamical variables, their action is in general dissipative.

In case of the Einstein-de Broglie theory of light, the loss of energy necessary to build Maxwell displacement current $\sim \frac{\partial \vec{E}}{\partial t}$ can only be borrowed from the real photon and associated electromagnetic field energy i.e., from the photon energy since they carry all observable energy independently of the surrounding electromagnetic wave's amplitude, as shown by Einstein in his theory of photoelectric effect.

(ii) In a recent work, we [Roy, 1996] have shown that this dissipation can be related to the fluctuation of the medium which can be characterized by random refractive index.

This can be envisaged as follows : Starting with the equation (5.54) and taking the time dependent part of $\vec{E}$, we get

$$\nabla^2 \vec{E} + (\chi^2 k_0^2 \chi_m \chi_e + i\mu_0 \chi_m \sigma k_0) vecE = 0 \tag{5.99}$$

Now

$$\eta = (\mu_0 \epsilon_0)^{\frac{1}{2}}; \quad \text{or} \quad \mu_0 = \frac{\eta^2}{\epsilon_0}$$

So,

$$\nabla^2 \vec{E} + \eta^2 k_0^2 \chi_m \chi_e \vec{E} + i\eta^2 \frac{\chi_m k_0 \sigma}{\epsilon_0} \vec{E} = 0 \tag{5.100}$$

This is a random differential equation having the random coefficient of $\vec{E}$. Several authors studied the wave equation with random refractive index [Frisch 1964] in the form

$$\nabla^2 \Psi + k_0^2 (1 + \mu(\omega, r)) \Psi = 0 \tag{5.101}$$

in which $\eta^2 = 1 + \mu(\omega, r)$ is the random refractive index. By replacing k_0 by $k_0 + i\eta (\eta > 0)$, it has been shown [Frisch, 1964] that the above equation [with Ψ] has bounded and unique solution in the asymptotic region (i.e., $r \to \infty$). This holds for small values of η i.e., $\eta \to 0$. Then the above equation can be written as

$$\nabla^2 \Psi + k_0^2 (1 + \mu) \Psi + 2i\eta k_0 (1 + \mu) \Psi = 0 \tag{5.102}$$

Comparing this equation with the above equation for $\vec{E}$ we can have

$$\eta \simeq \frac{\chi_m \sigma}{\epsilon_0}$$

Since the bounded solution exists for $\eta \to 0$, σ should be very small. Hence the equation(5.102) has a bounded and unique solution in the asymptotic region for $\sigma \to 0$. I n fact, the introduction of $k_0 + i\eta$ instead of k_0 implies the introduction of small dissipation during the propagation of a wave in a medium with random refractive index. So we can associate the dissipation of the energy of the photon in a Maxwell vacuum ($\sigma \neq 0$) with a fluctuating vacuum which can be characterized by a random refractive index.

In our framework of Maxwell vacuum [Roy 1996] with $\sigma \neq 0$, a dispersion relation can be written in a covariant form as

$$[|\vec{k}|^2 - m^2 k_0^2] A^\mu(k) = \frac{\mu}{c}\left(g^{\mu\nu} + \frac{n^2-1}{n^2} u^\mu u^\nu\right) J_\nu(k) \tag{5.103}$$

with $k = (\vec{k}, k_0)$, $J^\mu = (0, \sigma\vec{E})$, $A^\mu = (\vec{A}, i\phi/c)$.

Here we have taken $< \frac{1}{\eta^2} >= \frac{1}{n^2}$. $<,>$ being the statistical average of the random refractive index. So, we can think of effective metric tensor of Maxwell vacuum $G_{\mu\nu}$ as

$$G^{\mu\nu} = g^{\mu\nu} + \frac{n^2-1}{n^2} u^\mu u^\nu \tag{5.104}$$

by taking the average of random refractive index.

This form of the metric tensor has been already discussed by Synge et al. [Synge 1966] in studying the gravitational and electromagnetic fields on the generalized Lagrange space in dispersive as well as in non-dispersive medium from the point of view of relativistic geometrical optics. They have considered a generalized metric tensor $g_{\mu\nu}(x, V(x))$ as

$$g_{\mu\nu}(x, V(x)) = \Gamma_{\mu\nu}(x) + (1 - \frac{1}{n^2(x, V(x))}) V_\mu V_\nu \tag{5.105}$$

where $V_\mu(x)$ is the velocity field of the medium. The medium $\mathrm{M} = [M, V(x), n(x, V(x))]$ is called the dispersive medium, M being the manifold and $n(x, V(x))$ the refractive index.

If $\frac{\partial n}{\partial V} = 0$, then $\mathcal{M}$ is called a non-dispersive medium

Again if $\frac{1}{n^2} = 1 - \frac{1}{c^2}$, then $g_{\mu\nu}(n, V(x))$ is reduced to the metric $(T_{\mu\nu} + \frac{1}{c} Y_\mu Y_\nu)$ which was considered by Miron and Kawaguchi[Miron 1991].

It is easy to check that

$$g_{\mu\nu}(x, V) G^{\nu\xi}(x, V) = \delta^\xi_\mu \tag{5.106}$$

In our framework, the metric tensor constructed by averaging over the refractive index should be metric tensor for non-dispersive medium i.e, $\frac{\partial n}{\partial V} = 0$. So, the metric of Maxwell vacuum with $\sigma \neq 0$ should be Finslerian in nature. But for $n = 1$, we get the usual Maxwell vacuum where the photon does not loose energy during its propagation.

(iii) The mechanism of dissipation should be related to general theory of relativity. If we consider interactions between Dirac's nonzero mass vacuum elements (nonzero photon) and heavier particles (waves) moving in such a stochastic medium, general relativity implies action and reaction between the $g_{\mu\nu}$ field and any moving object characterized by an energy momentum distribution $T_{\mu\nu}$ since

$$\begin{aligned} G_{\mu\nu} &= R_{\mu\nu} - \tfrac{1}{2} G_{\mu\nu} R + \Lambda G_{\mu\nu} \\ &= 8\pi G (T_{\mu\nu} + g_{\mu\nu} \rho_{\text{vac}}) \end{aligned} \tag{5.107}$$

This indicates that the motion of light is modified by the surrounding $g_{\mu\nu}$ field, but vice-versa the passage of a light pulse modifies the surrounding $g_{\mu\nu}$ field. This effect, is, of course, very small and has been generally neglected in the literature.

This is a variant of famous tired photon hypothesis. The cosmological significance of this effect will be discussed in the next chapter.

"When there was neither aught nor naught, when darkness was rolling over darkness, what existed ? and the answer is given - It then existed without vibration ... Then Kalpa begins, the Anidavatam (means existed without vibration) commence to vibrate and blow after blow is given by Prana to Akasha"

RIG - VEDA : Hymn of Creation.

6. EINSTEIN EQUATION FOR GRAVITATIONAL FIELD :

The metric tensor might be a random quantity in those regions of space where there are powerful turbulent motion of matter accompanied by considerable changes in the density of matter or having large irregular velocities of motion. This means that material energy-momentum tensor $T_{\mu\nu}$ is a random function. Therefore, it follows from the Einstein field equations that the metric tensor must also manifest stochastic fluctuations.

6.1 Einstein Equations and Fluctuation of $g_{\mu\nu}$:

(a) Rosen's Approach

Einstein equations can be written as [Rosen 1980]

$$\begin{aligned} R_{\mu\nu} &= -8\pi G(T_{\mu\nu} - \frac{1}{2}g_{\mu\nu}g^{\rho\sigma}T_{\rho\sigma}) \\ &= -8\pi G T^{*}{}_{\mu\nu} \end{aligned} \tag{6.1}$$

for each space-time cosmology in a Gibbsian statistical ensemble. In terms of the affine connection, the Ricci tensor on the left hand side is

$$R_{\mu\nu} = \Gamma^{\lambda}_{\mu\lambda,\nu} - \Gamma^{\lambda}_{\mu\nu,\lambda} + \Gamma^{\lambda}_{\mu\beta}\Gamma^{\beta}_{\nu\lambda} - \Gamma^{\alpha}_{\mu\beta}\Gamma^{\beta}_{\alpha\nu} \tag{6.2}$$

while the affine connection itself is defined in terms of the metric tensor by

$$\Gamma^{\rho}_{\mu\nu} = \frac{1}{2} g^{\rho\sigma}(g_{\sigma\mu,\nu} + g_{\sigma\nu,\mu} - g_{\mu\nu,\sigma}) \tag{6.3}$$

We shall use the brackets to denote average values with respect to the statistical ensemble and putting $\overline{g}_{\mu\nu} =< g_{\mu\nu} >$, $\overline{\Gamma}^{\rho}_{\mu\nu} =< \Gamma^{\rho}_{\mu\nu} >$, we have

$$\begin{aligned} < R_{\mu\nu} > &= \overline{\Gamma}^{\alpha}_{\mu\alpha,\nu} - \overline{\Gamma}^{\alpha}_{\mu\nu,\alpha} + \overline{\Gamma}^{\alpha}_{\mu\beta}\overline{\Gamma}^{\beta}_{\nu\alpha} - \overline{\Gamma}^{\alpha}_{\mu\nu}\overline{\Gamma}^{\beta}_{\alpha\beta} \\ &+ < \gamma^{\alpha}_{\mu\beta}\gamma^{\beta}_{\nu\alpha} > - < \gamma^{\alpha}_{\mu\alpha}\gamma^{\beta}_{\nu\beta} > \end{aligned} \tag{6.4}$$

and

$$< g_{\rho\sigma}\Gamma^{\rho}_{\mu\nu} >= \frac{1}{2}(\overline{g}_{\sigma\mu,\nu} + \overline{g}_{\sigma\nu,\mu} - \overline{g}_{\mu\nu,\sigma}) \tag{6.5}$$

The tensor

$$\gamma^{\rho}_{\mu\nu} = \Gamma^{\rho}_{\mu\nu} - \overline{\Gamma}^{\rho}_{\mu\nu}$$

relates fluctuations in the gravitational force field with zero mean value $< \gamma^{\rho}_{\mu\nu} >= 0$, as a consequence of the deviation

$$\overline{\Gamma}^{\rho}_{\mu\nu} =< \Gamma^{\rho}_{\mu\nu} > \quad \text{from} \quad \Gamma^{\rho}_{\mu\nu}$$

The probability distribution over cosmologies is Riemannian if at every space-time point

$$< (g_{\sigma\rho} - \overline{g}_{\sigma\rho})\gamma^{\rho}_{\mu\nu} >= 0 \tag{6.6}$$

Then the mean value of the affine connection is given by the Christoffel symbols in mean value of the metric tensor

$$\overline{\Gamma}^{\rho}_{\mu\nu} = \left\{ \begin{matrix} \rho \\ \mu\nu \end{matrix} \right\} = \frac{1}{2}\overline{g}^{\rho\sigma}(\overline{g}_{\sigma\mu,\nu} + \overline{g}_{\sigma\nu,\mu} - \overline{g}_{\mu\nu,\sigma}) \tag{6.7}$$

Now the Einstein equations for the mean value of the metric tensor can be written as

$$\left\{ \begin{matrix} \alpha \\ \mu\alpha \end{matrix} \right\}_{\rho\nu} - \left\{ \begin{matrix} \alpha \\ \mu\nu \end{matrix} \right\}_{\rho\alpha} + \left\{ \begin{matrix} \alpha \\ \mu\beta \end{matrix} \right\}\left\{ \begin{matrix} \beta \\ \nu\alpha \end{matrix} \right\} - \left\{ \begin{matrix} \alpha \\ \mu\nu \end{matrix} \right\}\left\{ \begin{matrix} \beta \\ \alpha\beta \end{matrix} \right\} = -9\pi G\overline{T^{*}_{\mu\nu}} \tag{6.8}$$

where

$$\overline{T^{*}_{\mu\nu}} =< T^{*}_{\mu\nu} + (8\pi G)^{-1} + (\gamma^{\alpha}_{\mu\beta}\gamma^{\beta}_{\nu\alpha} - \gamma^{\alpha}_{\mu\nu}\gamma^{\beta}_{\alpha\beta}) > \tag{6.9}$$

The term $\overline{T}^{*}_{\mu\nu}$ is known as effective source term. This appears in Einstein equations if the probability distribution is Riemannian in the sense defined above.

Since all of the homogeneous and isotropic cosmological models of the universe can be cast in conformally Minkowskian form, it is reasonable to consider conformally related metric tensors as

$$g_{\mu\nu} = \exp[\phi]\overline{g}_{\mu\nu} \tag{6.10}$$

in which ϕ is a stochastic scalar field. We assume

$$\overline{g}_{\mu\nu} =< g_{\mu\nu} > \tag{6.11}$$

which means $< \exp[\phi] >= 1$ at all space-time points. The normalization condition imply that

$$< \exp[\phi]\phi_{,\mu} >= 0 \tag{6.12}$$

Then

$$\Gamma^{\rho}_{\mu\nu} = \left\{ {\rho \atop \mu\nu} \right\} + \frac{1}{2}(\phi_{,\mu}\delta^{\rho}_{\nu} + \phi_{,\nu}\delta^{\rho}_{\mu} - g_{\mu\nu}g^{\rho\sigma}\phi_{,\sigma}) \tag{6.13}$$

Hence the probability distribution is Riemannian with $\overline{T}^{\rho}_{\mu\nu} = \left\{ {\rho \atop \mu\nu} \right\}$ if and only if the homogeneity condition $< \phi_{,\mu} >= 0$ is satisfied.

Now the fluctuations in the gravitational force field can be written as

$$\gamma^{\rho}_{\mu\nu} = \frac{1}{2}(\phi_{,\mu}\delta^{\rho}_{\nu} + \phi_{,\nu}\delta^{\rho}_{\mu} - g_{\mu\nu}g^{\rho\sigma}\phi_{,\sigma}) \tag{6.14}$$

and the mean value of the stochastic stress-energy tensor is assumed to take the generic form

$$< T_{\mu\nu} >\simeq< \rho + p > u_{\mu}u_{\nu}+ < p > \overline{g}_{\mu\nu} \tag{6.15}$$

Here, u_{μ} is the normalized time like eigenvector of $< T_{\mu\nu} >$, possibly but not necessarily the local four velocity for galactic motion ($u_{\rho}u^{\rho} = -1$). So

$$\begin{aligned} < T^{*}_{\mu\nu} > &= < T_{\mu\nu} > -\frac{1}{2}g_{\mu\nu}g^{\rho\sigma} < T_{\rho\sigma} > \\ &\simeq < \rho + p > u_{\mu}u_{\nu} + \frac{1}{2}(\rho - p)\overline{g}_{\mu\nu} \end{aligned} \tag{6.16}$$

Since the correlation tensor $< \phi_{,\mu}\phi_{,\nu} >$ is nonnegative definite and the only preferred direction at a space-time point is given by time like u_{μ}, we must have

$$< \phi_{,\mu}\phi_{,\nu} >= A(u_{\mu}u_{\nu} + \xi\overline{g}_{\mu\nu}) \tag{6.17}$$

for certain nonnegative scalar functions A and $\xi(<1)$. Therefore, it follows that

$$\begin{aligned} <\gamma^\alpha_{\mu\beta}\gamma^\beta_{\nu\alpha}-\gamma^\alpha_{\mu\nu}\gamma^\beta_{\alpha\beta}> &= \frac{1}{2}[\overline{g}_{\mu\nu}<\phi_{,\alpha}\phi^\alpha_{,}>-<\phi_{,\mu}\phi_{,\nu}>] \\ &= \frac{1}{2}A[(3\xi-1)\overline{g}_{\mu\nu}-u_\mu u_\nu] \end{aligned} \tag{6.18}$$

Hence, the effective stress energy tensor can be written as

$$\begin{aligned} \overline{T}^*_{\mu\nu} &\simeq [<\rho>+<p>-(16\pi G)^{-1}A]u_\mu u_\nu \\ &\quad +\frac{1}{2}[<\rho>-<p>+(8\pi G)^{-1}(3\xi-1)A]\overline{g}_{\mu\nu} \\ &= (\overline{\rho}+\overline{p})u_\mu u_\nu+\frac{1}{2}(\overline{\rho}-\overline{p})\overline{g}_{\mu\nu} \end{aligned} \tag{6.19}$$

where the effective density and pressure are

$$\overline{\rho}=<\rho>+3(16\pi G)^{-1}(\xi-\frac{1}{2})A \tag{6.20}$$

$$\overline{p}=<p>+(16\pi G)^{-1}(\frac{1}{2}-3\xi)A \tag{6.21}$$

(b) Blokhintsev framework

However, Blokhintsev [Blokhintsev 1975] approached the problem in a different way. In this case, energy-momentum tensor $T_{\mu\nu}$ is also a random function but this can be expressed explicitly by introducing the stochastic parameter ζ, i.e.,

$$T_{\mu\nu}=T_{\mu\nu}(x,\zeta) \tag{6.22}$$

These random variables can be the initial values of the Lagrangian coordinates of the particles.

From the Einstein equation

$$R_{\mu\nu}-\frac{1}{2}g_{\mu\nu}R=8\pi GT_{\mu\nu}(x,\zeta) \tag{6.23}$$

it follows that the metric tensor $g_{\mu\nu}(x)$ is also a stochastic quantity which can be written as :

$$g_{\mu\nu}=g_{\mu\nu}(x,\zeta) \tag{6.24}$$

We shall assume that the fluctuations in the turbulent motion of the matter are small in comparison with the mean values of the characteristic quantities. Then it is possible to resolve the tensor into two parts

$$T_{\mu\nu} = \overline{T}_{\mu\nu}(x) + \Phi_{\mu\nu}(x,\zeta) \tag{6.25}$$

$\overline{T}_{\mu\nu}(x)$ being the mean values of the energy-momentum tensor

$$\overline{T}_{\mu\nu}(x) = \int T_{\mu\nu}(x,\zeta)d\omega(\zeta) \tag{6.26}$$

where $d\omega(\zeta)$ is the normalized probability of some particular distribution of the random parameters ζ. The tensor $\Phi_{\mu\nu}(x,\zeta)$ is determined entirely by the fluctuations of the moving matter. This tensor is assumed to be proportional to a suitable parameter ϵ that determines the amplitude of the fluctuations. Similarly, the metric tensor can be resolved as

$$g_{\mu\nu} = \overline{g}_{\mu\nu}(x) + h_{\mu\nu}(x,\zeta) \tag{6.27}$$

$\overline{g}_{\mu\nu}(x)$ is determined by the average motion of matter, while the tensor $h_{\mu\nu}(x,\zeta)$ is determined by the fluctuations of the matter and is also assumed to be proportional to the parameter ϵ. By equating the coefficients of the powers of ϵ in Einstein equation, we get

$$\overline{R}_{\mu\nu} - (\frac{1}{2})\overline{g}_{\mu\nu}R = (8\pi G)\overline{T}_{\mu\nu}(x) \tag{6.28}$$

and

$$A^{\rho\sigma}_{\mu\nu}h_{\rho\sigma} - B^{\rho\sigma\alpha}\frac{\partial h_{\rho\sigma}}{\partial x^\alpha} - C^{\rho\sigma\alpha\beta}_{\mu\nu}\frac{\partial h_{\rho\sigma}}{\partial x^\alpha \partial x^\beta} = 8\pi G\phi_{\mu\nu}(x,\zeta) \tag{6.29}$$

where

$$A^{\rho\sigma}_{\mu\nu} = \frac{\partial \overline{L}_{\mu\nu}}{\partial g_{\rho\sigma}}; \quad B^{\rho\sigma\alpha}_{\mu\nu} \equiv \frac{\partial \overline{L}_{\mu\nu}}{[\partial(\frac{\partial g_{\rho\sigma}}{\partial x^\alpha})]}$$

$$C^{\rho\sigma\alpha\beta}_{\mu\nu} = \partial\overline{L}_{\mu\nu}\partial\left(\frac{\partial^2 g^{\rho\sigma}}{\partial x^\alpha \partial x^\beta}\right)$$

and

$$\overline{L}_{\mu\nu} = \overline{R}_{\mu\nu} - (1/2)\overline{g}_{\mu\nu}R \tag{6.30}$$

It is to be noted that the coefficients A, B and C are of the same order of magnitude as $\overline{L}_{\mu\nu}$ i.e.,

$$\overline{L}_{\mu\nu} \simeq \frac{1}{l^2} \tag{6.31}$$

where l is the length scale determining the curvature of the space, whose metric is governed by the average motion. Again

$$8\pi G\overline{T}_{\mu\nu} = (8\pi G)\,\overline{\rho}\delta_{\mu\nu}$$

choosing the unit $c = 1$.

In this case, $\overline{\rho}$ is the average density of matter. This can be represented in the form

$$\bar{\rho} = \frac{\mu}{a^3} \tag{6.32}$$

with μ as the characteristic mass and a as the characteristic dimension of the system under consideration. Then after simple calculation one gets

$$1/l^2 \approx (a_g/a^3) \tag{6.33}$$

where a_g is the gravitational radius corresponding to the mass of the system. Consider the length scale characterizing the gradient of the stochastic field to be l'. Let the mass characteristic of the tensor fluctuation scale be Δm and b be the length determining the dimension of these fluctuations. Then the gravitational radius of the fluctuation $b_g = 2\pi G\Delta m$. and it is obvious that $l > l'$.
Again by virtue of the linearity of the equation for the tensor $h_{\mu\nu}$, the length l' is of the order of b. Thus

$$h \sim (l'^2/b^2)b_g = b_g/b \tag{6.34}$$

So the value of h will be very small if b_g/b is small.
Hence this quantity can be used as a small parameter ϵ in terms of which the stochastic tensors like $T_{\mu\nu}$ & $g_{\mu\nu}$ can be expanded.

Blokhintsev considered gravitational field fluctuation due to vacuum oscillations of matter. In vacuum

$$\overline{T}_{\mu\nu}(x) = 0 \tag{6.35}$$

Now the tensor $T_{\mu\nu} = \Phi_{\mu\nu}(\alpha, \zeta)$ and it describes the total fluctuations of all the fields with allowance for their interaction. It can be calculated for different types of fields like scalar field, spinor field etc. The mean value of the metric tensor components are $\bar{g}_{\mu\nu} = g^0_{\mu\nu}$, where $g^0_{\mu\nu}$ is the value of this tensor for a pseudo-Euclidean space. The equation (6.28) takes the simple form

$$\frac{1}{2}\Box\psi^{\mu}_{\nu} = -(8\pi G)T^{\mu}_{\nu} \tag{6.36}$$

and

$$\partial\psi^{\mu}_{\nu}\partial x^{\mu} = 0 \tag{6.37}$$

The temsor $T_{\mu\nu}$ describes the total fluctuations of all the fields with allowance for their interaction. The following calculations indicate that the conclusion do not depend significantly on the particular form of the field.

In case of a scalar field ϕ the Langrangian is

$$L = \frac{1}{2}(g^{\alpha\beta}\frac{\partial\phi}{\partial x^{\alpha}}\frac{\partial\phi}{\partial x^{\beta}} - m^2\phi^2) \tag{6.38}$$

and the corresponding contravariant momentum energy tensor is

$$T_{\mu\nu} = g^{\mu\nu}g^{\nu\beta}\frac{\partial\phi}{\partial x^{\alpha}}\frac{\partial\phi}{\partial x^{\beta}} - g^{\mu\nu}L \tag{6.39}$$

ϕ can be represented by a series in terms of creation and annhilation operators. They satisfy the commutation relation

$$[\hat{a}_k, \hat{a}^{+}_{k'}] = \delta_{kk'} \tag{6.40}$$

In the case of spinor field ψ we have

$$T_{\mu\nu} = \frac{1}{2}(\bar{\psi}\gamma^{\nu}\frac{\partial\psi}{\partial x^{\mu}} - \frac{\partial\bar{\psi}}{\partial x^{\mu}}\gamma^{\nu}\psi) \tag{6.41}$$

Similar to scalar field ϕ, ψ and its conjugate field can be also expanded in terms of creation and annihilation operators. Now according to the definition of the normal product, the mean value of $T_{\mu\nu}$ over the vacuum is zero. From (6.36) we get

$$\psi^{\mu}_{\nu}(x) = A\int G(x - x')\hat{T}^{\mu}_{\nu}(x')d^3x' \tag{6.42}$$

where $A = 16G$, G being Gravitational constant. $G(x - x')$ is the Green's function of the free equation. The correlation of the quantities $\psi^{\mu}_{\nu}(x)$ and $\psi^{\alpha}_{\beta}(y)$ can be calculated as

$$\overline{\psi^{\mu}_{\nu}(x)\psi^{\alpha}_{\beta}(y)} = A^2\int G(x - x')G(y - y')Q^{\mu\alpha}_{\nu\beta}(x' - y')d^4x'd^4y' \tag{6.43}$$

where the tensor $Q^{\mu\alpha}_{\nu\beta}$ is the correlation, given by

$$Q^{\mu\alpha}_{\nu\beta}(x-y) = \overline{T^{\mu}_{\nu}(x)T^{\alpha}_{\beta}(y)} \tag{6.44}$$

This correlation depends only on the difference $(x-y)$ as the vacuum is considered as homogeneous. We then obtain for the spectral expension of the correlation of the metric-tensor components

$$\begin{aligned} M^{\mu\alpha}_{\nu\beta}(x-y) &= \psi^{\mu}_{\nu}(x)\psi^{\alpha}_{\beta}(y) \\ &= \int M^{\mu\alpha}_{\nu\beta}(q)\exp[iq(x-y)]d^4q \end{aligned}$$

Then

$$M^{\mu\alpha}_{\nu\beta} = A^2.\frac{1}{q^4}.Q^{\mu\alpha}_{\nu\beta}(q) \tag{6.45}$$

The factor q^4 is due to

$$G(q) \sim \frac{1}{q^2} \tag{6.46}$$

The correlation $Q^{\mu\alpha}_{\nu\beta}$ can be calculated for scalar and spinor fields.

Scalar Field :

Let us consider now the case of Scalar field which is as follows :

$$Q^{\mu\alpha}_{\nu\beta}(x-y) = \frac{1}{2}\left\{\overline{T^{\mu}_{\nu}(x)T^{\alpha}_{\beta}(y)} - \overline{T^{\mu}_{\nu}(y)T^{\alpha}_{\beta}(x)}\right\}$$

Here, we have considered symmetrized correlation. After some straightforward calculations, we obtain

$$Q^{\mu\alpha}_{\nu\beta}(r) = \frac{2}{(2\pi)^4}.\frac{\hbar^2k^4}{c^2}.\frac{1}{r^4}(1-\cos 2kr) \tag{6.47}$$

as $r = |x-y| \to \infty$ and $t \to 0$. For $t \to \infty$ and $r \to 0$, we obtain

$$Q^{\mu\alpha}_{\nu\beta}(t) = -\frac{1}{2(2\pi)^4}.\frac{\hbar^2k^6}{c^2}.\frac{1}{t^2}\cos 2kt \tag{6.48}$$

Then

$$Q^{00}_{00}(r) \sim \frac{\hbar^2}{c^2\Lambda^2} \tag{6.49}$$

where $\Lambda = \frac{2\pi}{k}$.

Consequently the amplitude of the oscillations of the density of scalar matter is

$$\Delta\rho \sim \hbar c(c^2\Lambda^4) \tag{6.50}$$

The ratio of gravitational radius of this density to the dimension of the fluctuation Λ is

$$\frac{b_g}{\Lambda} = \frac{\Lambda_g^2}{\Lambda^2} \tag{6.51}$$

Therefore the gravitational effects due to vacuum fluctuations of the scalar fields with linear scalar Λ do not depend on the particle mass if $\Lambda \ll \frac{\hbar}{mc}$ and are small so long as

$$\Lambda > \Lambda_g \simeq 10^{-32} \text{ cm} \tag{6.52}$$

Spinor Field :

Performing calculations analogous to that of scalar field, we obtain

$$\begin{aligned} Q^{00}_{00}(x-y) &= \frac{1}{4}\cdot\frac{1}{2\pi^4}\cdot\frac{k^2}{c^2}\int d^3k d^3k'(\omega_k - \omega_{k'})^2 \\ &\times \; M^2(k,k')\cos((k+k'),(x-y)) \end{aligned} \tag{6.53}$$

with $M^2(k,k') = 1$.

For $r \to \infty$ and $t = 0$,

$$Q^{00}_{00} = \frac{1}{(2\pi)^4}\cdot\frac{\hbar^2}{c^2}\cdot\frac{k^2m^2}{c^2}\cdot\frac{1}{r^4}(1 - \cos 2kr) \tag{6.54}$$

At $t \to \infty$ and $r = 0$

$$Q^{00}_{00} = \frac{1}{2\pi^4}\cdot\frac{\hbar^2k^4m^2}{c^2}.\cos 2kt \tag{6.55}$$

Here, m is the mass of the spinor.

It is now evident from both these cases that the density amplitude $\Delta\rho$ is particularly the same in both the cases and does not depend on the mass of the particle. On the otherhand the behavior of large r or t is somewhat different in the case of Bose statistics and Fermi statistics. This should be studied in great details so as to understand the space-time structure and quantum statistics. It may give new insight on the behavior of microparticles at low temerature. But he did not elaborate the cosmological scenario within his framework. It requires serious efforts to understand the problems arising out of the modern debate on cosmology. On the otherhand, within Rosen's framework it is easier to have immediate insights on some cosmological issues.

(C) Nariai's Approach :

Nariai [Nariai, 1974, 1975] considered Einstein equations from the phenomenological point of view. The basic idea is that modified equations are constructed for mean value field of a highly supersonic turbulent gravitohydrodynamic field whose substratum is dominated by the hadronic matter with large density fluctuation i.e., Hagedron's hadronic matter based on the concept of statistical bootstrap. In consequence, there have been derived dynamical equations for mean value quantities of the turbulent field, which are greatly simplified when the mean-value metric represents the regular isotropic model universe.

The modified gravitational Lagrangian introduced phenomenologically can be written as

$$\begin{aligned} L_g &= R - l^{-2} f(l^2 R) \\ &= R\Big\{1 - 6x + 9x^{n-3} F(n, (n-4), (2n - 5/2); x)\Big\} \end{aligned}$$

where $x = (l^2 R)^{\frac{1}{3}}$ and $R = g^{ij} R_{ij}$, being the scalar curvature, F the hypergeometric function with $n = \frac{(7+\sqrt{73})}{4}$; $(2n^2 - 7n - 3 = 0)$.

The above gravitational Lagrangian leads to the following field equations

$$R_{ij} - \frac{1}{2} R g_{ij} - l^{-2}\Big\{\Big(\frac{\partial f}{\partial R}\Big) R_{ij} - \frac{1}{2} f g_{ij} - \Big(\frac{\partial f}{\partial R}\Big)_{;ij} + g_{ij}\Big(\frac{\partial f}{\partial R}\Big)^{;jk}_{;k}\Big\} = kT_{ij} \quad (6.56)$$

where R_{ij} is the Ricci tensor, $R = g^{ij} R_{ij}$, T_{ij} is the energy-momentum tensor and a semicolon stands for covariant differentiation. It follows from the above field equations that

$$T^{ij}_{;j} = 0 \quad (6.57)$$

as in general relativity.

The regular isotropic model universe cosisting of a hadronic substratum (specified by $\frac{p}{\rho c^2} = 0$ in a good approximation) is represented by the metric

$$ds^2 = -\bar{g}_{ij}dx^i dx^j = a^2(\eta)(d\eta^2 - \delta_{\alpha\beta}dx^\alpha dx^\beta) \tag{6.58}$$

Here $\bar{g}_{ig}$ is the mean metric for highly turbulent gravitational field. Let $\bar{g}_{ij}$ be the metric tensor for turbulent field. Then

$$g_{ij} = \bar{g}_{ij} + h_{ij} \tag{6.59}$$

and

$$g^{ij} = \bar{g}^{ij} + \hat{h}^{ij} \tag{6.60}$$

h_{ij} being the fluctuating part of g_{ij} whose mean value can be written as

$$< h_{ij}(t, \vec{x}) >= \frac{1}{V_i}\int_{V_i} h_{ij}(t + t', \vec{x} + \vec{x}')[-g(t', \vec{x}')]^{\frac{1}{2}} \tag{6.61}$$

in which V_i stands for a characteristic 4-dimensional volume in the turbulent space-time.Here, Einstein equations are assumed to be valid for fluctuating metric tensor g_{ij} i.e.,

$$G_i^j = R_i^j - \frac{1}{2}\delta_i^j R = kT_i^j \tag{6.62}$$

Similarly for mean value, we obtain

$$\bar{R}_i^j - \frac{1}{2}\delta_i^j \bar{R} + < \delta G_i^j - k\delta T_i^j >= k\bar{T}_i^j \tag{6.63}$$

where $< \delta G_i^j >=< \delta R_i^j > -\frac{1}{2}\delta_i^j < \delta R >$.

The main task of Nariai's approach is to look for a possibility of identifying equation (6.57) with equation (6.63) for the mean value field.

For this purpose, we must have

$$< \delta G_i^j > -k < \delta T_i^j >= l^{-2}\left[\left(\frac{\partial f}{\partial R}\right)^{;j}_{;i} - \left(\frac{\partial f}{\partial R}\right)R_i^j - \delta_i^j\{\left(\frac{\partial f}{\partial R}\right)^{;k}_{;jk} - \frac{1}{2}f\}\right] \tag{6.64}$$

whose contraction gives

$$< \delta G > -k < \delta T >= -l^{-2}\Big\{3\Big(\frac{\partial f}{\partial R}\Big)^{;k}_{;k} + R\left(\frac{\partial f}{\partial R}\right) - 2f\Big\} \tag{6.65}$$

Again we have

$$T^j_{i|j} \equiv T^j_{i,j} + \Gamma^j_{kj}T^k_i - \Gamma^k_{ij}T^j_k 00 \tag{6.66}$$

After some calculations this yields

$$(\delta T^j_i)_{;j} = \rho^k_{ij}(\bar{T}^j_k + \delta T^j_k) - \rho^j_{kj}(\bar{T}^k_i + \delta T^k_i) \tag{6.67}$$

Its mean value equation is of the form

$$< (\delta T^j_i)_{;j} >=< \rho^k_{ij} > \bar{T}^j_k - < \rho^j_{kj} > \bar{T}^k_i + < \rho^k_{ij}\delta T^j_k - \rho^j_{kj}\delta T^k_i > \tag{6.68}$$

where,

$$\begin{aligned} \rho^k_{ij} &= \Gamma^k_{ij} - \bar{\Gamma}^k_{ij} \\ \delta T^j_i &= T^j_i - \bar{T}^j_i \\ \delta G^j_i &= G^j_i - \bar{G}^j_i \end{aligned}$$

Let us now adopt the gauge condition

$$h_{00} = h_{0\alpha} = 0 \tag{6.69}$$

so that the remaining six components $h\beta_\alpha$ can be decomposed into the transverse traceless part $h^{\beta TT}_\alpha$ (gravitational wave), the rotational part $h^{\beta R}_\alpha$ (rotational wave) and the longitudinal part $h^{\beta L}_\alpha$ (density wave). Then the surviving components of h^{ij} satisfying is

$$\hat{h}^{ij} = -g^{ij}h^j_k \tag{6.70}$$

where $h^j_k \equiv \bar{g}^{ij}h_{ik}$ are given by

$$\hat{h}^{\alpha\beta} = -h^{\beta\lambda}(\delta^\alpha_\lambda + h^\alpha_\lambda)^{-1} \tag{6.71}$$

on the prescription that $|h^\alpha_\lambda| \leq 1$ and

$$\left\{(\delta^\alpha_\lambda + h^\alpha_\lambda)^{-1}\right\}_{;i} = -(\delta^\mu_\lambda + h^\mu_\lambda)^{-1}(h^\nu_\mu)_{;i}\ (\delta^\alpha_\mu + h^\alpha_\nu)^{-1} \tag{6.72}$$

This leads to

$$< \delta G > -k < \delta T >= 0 \tag{6.73}$$

For simplicity, let us assume that, similarl to $\bar{T}_i^j$ corresponding to $\bar{g}_{ij}$, the fluctuating energy momentum tensor T_i^j is also of the form for a perfect fluid i.e.

$$T_i^j = (\rho + p) g_{ik} u^k u^j + p\delta_i^j \tag{6.74}$$

where

$$\begin{aligned} \rho &= \bar{\rho} + \delta\rho \\ p &= \bar{p} + \delta p \\ \frac{p}{\rho} &= 0 \end{aligned}$$

$\bar{\rho}$ and $\bar{p}$ are mean density and mean pressure respectively. In this case, $u^i = (\frac{1}{a} + \delta u^0, u^\alpha)$ is the 4-velocity of the turbulent field such that

$$g_{ij} u^i u^j = -1 \tag{6.75}$$

Let us now rewrite the following equation

$$\bar{R}_i^j - \frac{1}{2}\delta_i^j \bar{R} = k\bar{S}_i^j \quad \text{and} \quad \bar{S}_i^j \equiv \bar{T}_i^j + < \delta T_i^j - \bar{k}\delta G_i^j > \tag{6.76}$$

$\bar{S}_i^j$ plays a role of the energy momentum tensor in the usual formulation of general relativity. Inserting equations (6.74) into (6.76), we obtain

$$\rho_{\text{eff}} = -S_0^0 = \rho_* x^3 (1 - x) \tag{6.77}$$

$$p_{\text{eff}} = \frac{s_a^a}{3} = -\rho_* \frac{x^4}{3} \tag{6.78}$$

This shows that the efective pressure is negative, contrary to the effective density. Thaking l as the scale characterizing the hadronic matter i.e.

$$l = \frac{\hbar}{m_\pi c} \sim 10^{-18}\text{cm}$$

$$\rho_* \sim 5 \times 10^{52} \text{gm/cm}^3$$

It is worth mentioning that Nariai himself pointed out two defects in his approach. In the above calculations, possible interaction among TT-mode, R-mode and L-mode of the turbulent field has been disregarded. Another is that only the second order approximation itself which is used to calculate the mean value quantity.

6.2 Turbulent Gravitational Field & Cosmological issues :

(a) Cosmological Constant Problem :

The introduction of cosmological constant in gravitation theory has been questioned from time to time since it was proposed by Einstein himself. The work of Zeldovich [Zeldovich 1968] created fresh interest when he tried to show that the quantum fluctuations in vacuum may give rise to a term of the form $g_{\mu\nu}$ in the stress energy tensor. However, the fundamental questions regarding the $\Lambda \neq 0$ value and its numerical estimate are not answered unambiguously even in this framework.

Recently Frieman [Frieman 1993] made a comprehensive review on recent developments in cosmology like the evidence for dark matter on large scales, inflationary models for the early universe, and constraints on models for large-scale structure formation from the recent COBE detection of microwave background anisotropy. The observations on CMBR and galaxy indicate that the homogeneous and isotropic FRW (Friedman-Robertson-Walker) models may be considered as a first approximation because the gravitational potential and thus the perturbation to FRW space-time metric, associated with large-scale inhomogeneties is relatively small, $\delta\phi \sim \delta g_{\mu\nu} \sim 10^{-5}$. The FRW models are characterized by a global scale factor $a(t)$, whose dynamics is determined by the matter content of the universe through Einstein's equations,

$$H^2 \equiv (\frac{\dot{a}}{a})^2 = \frac{8\pi G}{3}\rho - \frac{k}{a^2} + \frac{\Lambda}{3} \tag{6.79}$$

and

$$\frac{\ddot{a}}{a} = -\frac{4\pi G(\rho + 3p)}{3} + \frac{\Lambda}{3} \tag{6.80}$$

Here ρ is the mean energy density of matter, p is its pressure, Λ is the cosmological constant and the contribution from vacuum energy, $k = 0, 1, -1$ is the sign of the spatial curvature. Models with $k \leq 0$ are spatially infinite (open), while those with $k = 1$ are spatially finite (closed). From the Einstein equation, if $\Lambda = 0$ there is a one to one correspondence between the spatial geometry and the fate of the universe : open models expand forever, while closed models eventually recollapse. Astronomical observations clearly suggest that the non-vacuum energy density of the universe is currently dominated by non-relativistic matter, while the early universe was dominated by ultrarelativistic particles or radiation.

In FRW models, there are several observable cosmological parameters such as Hubble parameter, $H_0 = (\dot{a}/a)_0$ (the subscript 0 denotes the present epoch), the age of the universe, the present mass density relative to the 'critical density' of the spatially flat Einstein de Sitter ($K = \Lambda = 0$) model,

$$\Omega_0 = \rho_0/\rho_{\text{critcal}} = \frac{8\pi G\rho_0}{3H_0^2}$$

the deceleration parameter, $q_0 = -(\frac{\ddot{a}}{\dot{a}^2})_0$, which measures the rate at which the gravitational attraction of the matter is slowing down the expansion, and the contribution of the cosmological constant to the present expansion rate, $\lambda_0 = \frac{\Lambda}{3H_0^2}$. From the Einstein equations, these parameters are related by

$$\begin{aligned} 1 &= \Omega_0 + \lambda_0 - \frac{k}{a_0^2 H_0^2} \\ q_0 &= \frac{\Omega_0}{2} - \lambda_0 \end{aligned} \tag{6.81}$$

In addition, the age of the universe is related to the other parameters through an expression of the form

$$H_0 t_0 = f(\Omega_0, \lambda_0) \tag{6.82}$$

For matter dominated universe with $\Lambda = 0$, f falls monotonically with increasing Ω_0 and the two useful limits are $f(0,0) = 1$ and $f(1,0) = 2/3$.

More generally, over the range $0 < \Omega_a \leq 1$, $k \leq 0$, an excellent approximation is [Carroll 1992],

$$H_0 t_0 \simeq \frac{2}{3} \sinh^{-1} \frac{(\sqrt{\frac{|(1-\Omega_a)|}{\Omega_a}})}{\sqrt{|(1-\Omega_a)|}} \tag{6.83}$$

where

$$\Omega_a = \Omega_0 - 0.3(\Omega_0 + \lambda_0) + 0.3$$

Much efforts have been spent to measure or to contain all these parameters. There is still debate about the reliability and interpretation of much data already gathered. Let us discuss Rosen's framework of conformally fluctuating metric from which we get the altered solutions to the Friedman equations which is as follows:

$$\frac{\ddot{a}}{a} = -4\pi G(\frac{1}{3} + \overline{\rho} + \overline{p}) \tag{6.84}$$

and

$$(\frac{\dot{a}}{a})^2 + \frac{k}{a^2} = \frac{8}{3}\pi G\overline{\rho} \tag{6.85}$$

Here,

$$\overline{\rho} = <\rho> + 3(16\pi G)^{-1}(\xi - \frac{1}{2})A$$

$$\overline{p} = <p> + (16\pi G)^{-1}(\frac{1}{2} - 3\xi)A$$

Then

$$(\frac{\dot{a}}{a})^2 + \frac{k}{a^2} = \frac{8}{3}\pi G <\rho> + \frac{1}{2}(\xi - \frac{1}{2})A \tag{6.86}$$

Taking, $4\pi <\rho> a^3 = D$ (constant), we have,

$$(\dot{a})^2 + k - \frac{1}{2}(\xi - \frac{1}{2})Aa^2 = \frac{2}{3}\pi G/a \tag{6.87}$$

or,

$$(\dot{a})^2 + k - \Lambda a^2 = \frac{2}{3}\pi \frac{G}{a} \tag{6.88}$$

where $\Lambda = \frac{1}{2}(\xi - \frac{1}{2})A$.

This equation is identical to the Eddington-Lamatire model if we put [Sinha 1992]

$$\Lambda = \frac{1}{2}(\xi - \frac{1}{2})A.$$

Now two conditions arise i.e.,

– **(a)** If $\xi > \frac{1}{2}$, $\overline{\rho}$ must be greater than $<\rho>$.

– **(b)** If $\frac{1}{2} \leq \xi < 1$, then closed universe ($k = H$) may nevertheless be obtained.

The condition $\xi > \frac{1}{2}$ indicates that a significant level of cosmic turbulence is present for this range of ξ. It is evident from the above analysis that the metrical fluctuations always act to accelerate universal expansion, the non-negative incremental change in $\dot{a}/a$ being given as $\frac{1}{2}\xi A$. Again if $A = 0$, then $<\phi_{,\mu}\phi_{,\nu}> = 0$ so that there will be no stochasticity in the metric and $\Lambda = 0$. This added expansion rate due to metrical fluctuation may play important role in resolving the recent debate over the age of the universe and measured Hubble parameter.

So if we take the stochasticity of the space-time metric as an apriori notion, then obviously we reach at $\Lambda \neq 0$, which denotes the empty space producing the same gravitational field as when the space contains matter with mass density $\rho_\Lambda = \epsilon_\Lambda/c^2$ and the energy density $\epsilon_\Lambda = \frac{c^2\Lambda}{8\pi G}$ and pressure $p_\Lambda =$

$-\epsilon_A$. In this sense we can speak of an energy density of the vacuum. In fact, from different cosmological considerations and analysing observational data one can impose a limit

$$|\rho_\Lambda| < 5.10^{-29}\ gm/cm^2$$

corresponding to $|\Lambda| < 10^{-5} cm^{-2}$.

(b) Dominant Energy Condition :

Hawking [Hawking 1970] pointed out that the energy-momentum tensor will be said to satisfy the dominant energy condition if for every observer the local energy density T_{ab} is nonnegative and the local energy flow vector is non space-time where u_a is the velocity vector of the observer ($u^a u_a > 0$). This dominant energy condition can be equivalently written as

$$T_{00} \geq T_{ab} \quad \text{for each } a, b. \tag{6.89}$$

For a fluid with energy density ρ and principal pressure $p_i (i = 1, 2, 3)$ this will hold if and only if $\rho \geq |p_i|$.

Zeldovich and Pitaevskii [Zeldovich 1971] studied the possibility of violation of this dominant energy condition so as to explain the creation of particles by classical gravitational field. In the cases $k = 0$ and ± 1, for an arbitrary smooth function $R(g_{\mu\nu}), \overline{\rho}$ and $\overline{p}$ are shown to be finite over the interval $-\infty < t < +\infty$. Now, any smooth transition from contraction to expansion necessarily requires a point of maximal contraction somewhere in the interval. At this point, $R(t)$ has a minimum i.e.,

$$\frac{dR}{dt} = 0 \quad \text{and} \quad \frac{d^2R}{dt^2} = +ve \tag{6.90}$$

Hence, the sum $(\overline{\rho} + 3\overline{p})$ must be negative. Accordingly, if we assume that sum is always positive i.e., $(\overline{\rho} + 3\overline{p}) \geq 0$, then a smooth transition from contraction to expansion is impossible and a singularity $R \to 0$ is unavoidable. So, the dominant energy conditions will be violated if $(\overline{\rho} + 3\overline{p}) < 0$. It might be possible that vacuum energy density $\overline{p}$ is negative and $\overline{p}_{vac} = -\rho_{vac}$ is positive so that the effective pressure is greater than effective energy density i.e., $< \rho >= 0$ and zero average pressure i.e, $< p >= 0$ and zero tension. Then

$$\begin{aligned}\overline{\rho} + 3\overline{p} &= < \rho > +3 < p > - \frac{3A\xi}{8\pi G} \\ &= -\frac{3A\xi}{8\pi G}\end{aligned} \tag{6.91}$$

Since, A and ξ are two non-negative scalar functions,

$$\overline{\rho} + 3\overline{p} = -ve \tag{6.92}$$

for arbitrary A and $0 < \xi \leq 1$.

In this case

$$\frac{d^2R}{dt^2} = +ve$$

$$\frac{dR}{dt} = \frac{1}{2}(\xi - \frac{1}{2})A \quad \text{for } k = 0$$

$$\frac{dR}{dt} = 0 \quad \text{if } \xi = \frac{1}{2}$$

But for $\xi = \frac{1}{2}$, the cosmological constant

$$\Lambda = \frac{A}{2}(\xi - \frac{1}{2}) = 0.$$

So, the dominant energy condition will be violated for a vacuum with

$$< \rho >= 0, \ < p >= 0 \ \text{ and } \xi = \frac{1}{2} \tag{6.93}$$

In this case,

$$\overline{\rho} = 0$$

$$\overline{p} = \frac{1}{16\pi G}(\frac{1}{2} - 3\xi)A = -\frac{A}{16\pi G} \tag{6.94}$$

It indicates that dominant energy condition is violated under certain circumstances due to fluctuation of space-time metric. This is in accord with the arguments forwarded by Zeldovich and Pitaevskii that space-time curvature can first create pressure and tension and in turn the pressure and tension can cause the release of energy. The particle pair may be created due to this release of energy.

It may be pointed out that Nariai [Nariai 1974, 1975] also considered a highly turbulent Gravito-Hydrodynamic field in the hadron era of the universe. He also considered the fluctuation of the metric tensor due to presence of this turbulent field and calculated the effective density and pressure say ρ_{eff} and ρ_{eff} respectively. Within this framework,

$$\begin{aligned} \rho_{\text{eff}} &= \rho_* x^2(1 - x) \\ p_{\text{eff}} &= -\rho_* \frac{x^4}{3} \end{aligned} \tag{6.95}$$

which shows that the effective pressure is negative, contrary to the effective density. At the bounce epoch $x = 1$ in particular, we have

$$\rho_{\text{eff}} + 3p_{\text{eff}} = -\rho_A < 0$$

which means the violation of energy condition. Nariai has tried to construct a physical picture of this violation of energy condition. In the early universe when it was dominated by highly turbulent hadronic matter, the world line of matter at any space-time point was highly random contrary to the situation postulated in the proof of singularity theorems.

6.3 Possible Implications :

(i) Conformal Fluctuation of Metric

In chapter 2, we have already discussed the role of stochastic gravitational field in decoherence problem. Let us consider the stochastic fluctuation of conformal metric as described in (6.10) as follows :

$$g_{mu\nu} = \exp^{\phi(x)} \bar{g}_{\mu\nu}$$

where $\overline{g}_{\mu\nu} =< g_{\mu\nu} >$.

By assuming the probability distribution to be Riemannian, the following relations can be obtained,

$$< e^{\phi} >= 1; \quad < e^{\phi(x)}\phi_{,\mu} >= 0; \quad < \phi_{,\mu} >= 0$$

where the comma (,)indicates the derivatives with respect to spatial coordinates.

Now taking this conformally fluctuating metric as considered above, one can assume the correlation function as [Rosales 1995]

$$\begin{aligned} < \phi(x,t)\phi(x',t') > &= \sigma^2 \exp[-(t-t')^2/\tau^2] \\ &\quad \exp[-(x-x')^2/L^2] \end{aligned} \tag{6.96}$$

with $\tau = L/c$, L as the correlation length, $\sigma^2 \sim \hbar G/L^2c^2 = \Lambda_p^2/L^2$.

For all reasonable values of L, one has $\sigma << 1$, hence we are in the weak field case. So gravitational effects can be accounted for by stochastic Newtonian potential

$$\phi(x,t) \text{ and } V(x,t) = \frac{1}{2}mc^2\phi(x,t)$$

Diosi and Lukacs [Diosi 1993] pointed out some difficulties with respect to K-model [Karolyhazy] in connection with cosmological problems. Within the framework of K-model, it is predicted that there exists a cut-off parameter related to vacuum gravitational energy. However, this energy is larger than that of neutron star. Recently Rosales et al. [Rosales, 1995] claimed that no such problem arises in conformally fluctusting metric.In this approach, the mean value of space-time curvature can be written in terms of stochastic Newtonian potential $\phi(x,t)$ as

$$\overline{R} = -\frac{1}{2} < e^{-2\phi}\left(6\Box\phi + 3[(\frac{\partial\phi}{\partial t})^2 - (\nabla\phi)^2]\right) > \tag{6.97}$$

Using the statistical properties of stochastic scalar field $\phi(x,t)$, we get

$$< \Box\phi >= 0$$

$$< (\frac{\partial\phi}{\partial t})^2 >=< (\nabla\phi)^2 >= \frac{4\sigma^2}{L2} = \frac{4}{L^2}(\frac{\Lambda_p}{L})^2 \tag{6.98}$$

Then

$$\overline{R} = 0 \text{ up to order of } \phi^4.$$

Therefore, this model has problems as far as the mean curvature of the universe is concerned.

We should now compute the 00− component of the energy momentum tensor, which of course, is not a Minkowski scalar. However, the relative dynamic velocity (i.e., cosmic expansion not included) between any pair of physically relevant systems seems to be, as a matter of fact, much smaller than the velocity of light and hence one is allowed to assume $< T_{00} >$-component in the reference frame where

$$V(x,t) = \frac{1}{2}mc^2\phi(x,t)$$

holds. This gives a reliable estimate of the mean energy density of the universe stored in the fluctuating vacuum. Thus the mean energy density of the universe stored in the fluctuating vacuum becomes

$$< T_{00}/c^2 >= (c^2/32\pi GL^2)(\Lambda_p/L)^2 \tag{6.99}$$

Taking $\frac{<T_{00}>}{c^2} < 10^{-29}\text{gm/cm}^3$, one gets lower bound to L, i.e., $L \geq 10^{-3}$cm. Intuitively, one can think that the primordial gravity waves of the order of Planck length has "inflated" to the present order of L.

Another difficulty which arises in K-model i.e., metric fluctuation may induce oscillations of charged particles in a sample of condensed matter. This would spontaneously radiate a considerable amount of electromagnetic energy, mainly in the form of hard X-rays. But this does not happen. So let us discuss this problem in conformally fluctuating metric space. The dipole radiation formula can be usually written in the form [Landau 1962]

$$< W >= (\frac{2e^2}{3c^3}) < a^2 > \tag{6.100}$$

where, a is the stochastic acceleration of the particles by stochastic potential $V = \frac{1}{2}mc^2\phi(x,t)$. Then,

$$< W >= \frac{2e^2}{3c^3}.\frac{1}{m^2}.\frac{m^2c^4}{4} < (\nabla\phi)^2 >$$

$$< W >= \frac{2e^2}{3c^3}(\frac{c^2}{L})^2(\frac{\Lambda_p}{L})^2 \tag{6.101}$$

Taking $L \sim 10^{-3}$cm, we get,

$$< W >\simeq 5 \times 10^{-51} eV/\text{sec}$$

Then for a macroscopic system ($\sim 10^{23}$ particles), $< W >\approx 10^{-46}W$, which is absolutely negligible. So there is no such problem with the model which was present in K-model.

It should be mentioned that here in conformally fluctuating metric model L is not a cut-off parameter but a correlation length. For distances much less than it , metric fluctuations become rapidly correlated and this fact makes the reduction mechanism practically inefficient. The correlation length has been estimated and a lower bound of L can be of $\sim 10^{-3}$ cm. However, it should not be taken in a very strict sense. The value of the correlation length should be of the order of 10^{-3} cm means that there is a kind of "mesoscopic region", for masses about 10^{-10} gm, say, wherein some strange things could happen regarding the quantum seperation principle. This findings may play significant role in future research.

Wheeler, Gerlach and Dewitt [Wheeler 1968] in the theory of superspace initiated an investigation regarding the fluctuation of geometry and the structure of superspace. Here, the points are representatives of possible 3-geometries and in which the propagation of constructively interfering probability waves form a localized wave packet, the path of which marks the evolutionary trajectory of the corresponding classical $3D$-geometry in much the same way as the propagation of a localized solution of Schrödinger equation makes the path of the classical electron. The hope is that elementary particles can be described as quantum states of excitation of the geometry in terms of probability amplitudes which are functionals of possible 3D-geometries.

(ii) Woehler's Model

Woehler [Woehler 1971] tried to elaborate this idea and build up a model in which real turbulent fluctuations are superimposed on the average metric with the fluctuating metric satisfying the free space Einstein equations. This model is mainly based on Wheeler's conjecture that the quantum fluctuations of the metric create a multiply connected foam-like structure of the vacuum with a structure constant of

$$L^* = (\hbar G/c^3)^{1/2} \approx 10^{-33}\text{cm}$$

and large virtual energy densities of the order

$$c^4/GL^{*^2} \approx 10^{115}\text{eVg/m}^2$$

The basic assumption of this model is that the quantum fluctuation of the metric can be incorporated and it appears in the classical Einstein theory as a real random fluctuations of the metric field. There is no analog to this in classical field theory. Superposition of a random field in a linear theory does not produce any extra terms in the averaged field equations. Only in nonlinear theories, the introduction of turbulence may lead to new terms in the equation of motion like the Reynolds stresses in the Navier-Stokes equation of hydrodynamics [Woehler 1971]. But it is not clear in what sense such Reynold stresses might produce some semi-classical approximation to the correct quantum field theory. However, it should be noted that the superposition of a random field on the action function of classical mechanics does produce the correct field theory for the action function from which the Schrödinger equation is obtained.

The classical Hamilton-Jacobi equation for the action function $S(\vec{r}, t)$ can be written as

$$-\frac{\partial S}{\partial t} = \frac{(\nabla S)^2}{2m} + U(\vec{r}) \tag{6.102}$$

This can be interpreted in analogy to fluid dynamics as the equation of motion for a turbulent fluid with

$$S = S_0 + S_1$$

where S_0 is the average field and S_1 a random field with zero average. Putting S into the Hamilton-Jacobi equation and taking time average we obtain

$$-\frac{\partial S_0}{\partial t} = \frac{(\nabla S_0)^2}{2m} + U(\vec{r}) + \overline{(\nabla S_1)^2} \tag{6.103}$$

The last term on the right hand side of this equation may be thought as analog to the Reynolds stresses in the Navier-Stokes equations for turbulent flow. Now if we assume that the random field is stationary and the higher order autocorrelations are small, then one can show that the first nonzero term in the expansion for $\overline{(\nabla S_1)^2}$ would be $2\beta \nabla^2 S_0$ where β is a small constant. This extra term modifies Hamilton-Jacobi equation as

$$-\frac{\partial S_0}{\partial t} = \frac{(\nabla S_0)^2}{2m} + U(\vec{r}) + (\frac{\beta}{m})\nabla^2 S_0 \tag{6.104}$$

which is identical with the equation for the action function which one obtains from Schrödinger equation with

$$\Psi(\vec{r},t) = \exp[\frac{i}{\hbar}S(\vec{r},t)]$$

where β is taken to be $\beta = -\frac{i\hbar}{2}$.

The corresponding procedure for the Einstein theory would be to superimpose a random contribution on the Hamilton-Jacobi function in the Einstein-Hamilton-Jacobi equation of general relativity [Woehler, 1968]. Then one can solve the turbulence problem under suitable higher order moments. Here, one can get Reynold stresses in terms of the average action function. We can summerize the assumptions and postulates of this model as follows :

– **(a)** All physics is described by the metric field

$$g_{\mu\nu} = \bar{g}_{\mu\nu} + h_{\mu\nu} \tag{6.105}$$

where $t_{\mu\nu}$ is a random field in space and time and $\bar{g}_{\mu\nu}$ is an average field. The average of the random field is zero., i.e.,

$$< t_{\mu\nu} >= (\Delta V)^{-4} \int t_{\mu\nu}(x)(g)^{\frac{1}{2}} d^4x = 0 \tag{6.106}$$

– **(b)** The random field is assumed to have a spectral distribution which allows to decompose it into two parts

$$t_{\mu\nu} = f_{\mu\nu} + \phi_{\mu\nu}$$

$f_{\mu\nu}$ is associated with the vacuum fluctuations. It is postulated to be stationary, homogeneous, isotropic random field of normal distribution with $< f_{\mu\nu} >= 0$. The correlation length is taken to be of the order of L^* and the amplitude is taken to be of the order of $\bar{g}_{\mu\nu}$. The field $\phi_{\mu\nu}$ also has zero average but it is of long correlation length and is nonhomogeneous. It is zero for vacuum and where nonzero represents the excitons in terms of the correlation functions $< \phi_{\mu\nu}(x)\phi_{\delta\sigma}(x') >$ and higher order terms.

– **(c)** The amplitude $\phi_{\mu\nu}(x)$ is taken to represent a measure for the probability to find matter at point x.

– **(d)** The total field

$$g_{\mu\nu} = \bar{g}_{\mu\nu} + f_{\mu\nu} + \phi_{\mu\nu}$$

satisfies the free space Einstein equation given by

$$E_{\mu\nu} = R_{\mu\nu} - \frac{1}{2} g_{\mu\nu} R = 0$$

This is an expression of the assumption that matter is not some foreign agent in the arena of geometry but certain coharent wrinkles in the randomly fluctuationg geometry.

Ricci tensor $R_{\mu\nu}$, Christoffel symbol $\Gamma^{\rho}_{\mu\nu}$ scalar R all can be decomposed for fluctuating metric field. After taking averages we get modified Einstein equations. It contains two parts :

(i) Vacuum term $< V_{\mu\nu} >$
(ii) $< Y_{\mu\nu} >$ which contains the matter field $\phi_{\mu\nu}$ i.e.,

$$\bar{E}_{\mu\nu} =< V_{\mu\nu} > + < Y_{\mu\nu} > \tag{6.107}$$

where,

$$< V_{\mu\nu} <= \frac{1}{2} < (\bar{g}_{\mu\nu} + f_{\mu\nu}) f\delta R - f\delta R_{\mu\nu} >\equiv$$

$f\delta R_{\mu\nu}$ and $f\delta R$ contain nonlinear terms, the average of which do not vanish. Again,

$$< Y_{\mu\nu} >=< \frac{1}{2} g_{\mu\nu}\delta R + \delta R_{\mu\nu} - V_{\mu\nu} > \tag{6.108}$$

$< Y_{\mu\nu} >$ may be interpreted as the effective energy-momentum tensor multiplied with $\frac{8\pi G}{c^2}$, i.e.,

$$< Y_{\mu\nu} >= (\frac{8\pi G}{c^2})T_{\mu\nu}$$

where $T_{00} = \rho$ is the mass density.

It can be proved that in the limit of negligible source strength i.e., in a flat average metric, effective energy-momentum is conserved. The conservation equations

$$(\bar{g}^{\sigma\nu} < Y_{\mu\nu} >)_{;\delta} = 0$$

would be the field equations for matter. These are differential equations for the correlation function $\langle\phi_{\mu\nu}(x)\ \phi_{\mu\nu}(x') >$ and higher order moments. Stating from these equations a hierarchy of equations could be derived in the fashion of the turbulence theory of fluids. The solutions of these equations would have to represent the nature of possible excitons which can be described by these field equations. They have also to describe the lifetimes and the deacys into other excitons. They have to describe why some excitons are apparently stable and also the interactions of these excitons. it is very tmpting to find a rigorous formulations of excitons similar to the theory of elementary domains as developed by Yukawa school [Yukawa, 1950]. In this model, the role of universal radius R_0 raises some interest. The dependence of the exciton size and mass on this radius raises the questions of a possible time dependence of the fundamental particles. One possibility is to interpret R_0 as the radius of the universe at the phase of maximum expansion. Another possibility is to identify the length R_0 with the present average curvature of the universe. In such case the existence of the Hubble effect would force one to accept time dependence of the fundamental quantities in this model. The only length that appears explicitly in the equations by postulating the stationarity of the submicroscopic vacuum fluctuations in the length L^*. The radius R_0 which must enter in order to produce a length $L \sim 10^{-13}$ cm could enter in two ways : either through the average metric $\bar{g}_{\mu\nu}$ and the Christoffel symbols $\bar{\Gamma}^{\lambda}_{\mu\nu}$ or through boundary and initial conditions.

If R_0 determines the present properties of the exciton as a boundary condition representing the radius of a closed universe, then one would expect the exciton properties to be time-dependent togather with R_0. The time variability of the total exciton number N and the total mass M of the universe correspond to a mass production rate of 10^{-46}gm/cm^{3}sec which is similar to the rate in the C-field theory of Hoyle and Narlikar.

The time dependence of the exciton mass would imply a time dependence of the stellar masses and therefore, of the stellar luminosities. Estimates show that the luminosity for a main sequence star would depend on time like $\sim t^{-3}$ which is much weaker than the t^{-7} dependence in theories with variable gravitation constant. Corrections of the evolutionary age of the astrophysical objects due to the time dependence of the masses are similar to those given within the framework of Brans-Dicke Cosmology. The theory is far from complete and lot of efforts are needed to test the validity of this approach.

ORIGIN OF STOCHASTICITY

"The uncertainties of the world we now ascribe not to the uncertainties of our thoughts, but rather to the character of the world around us. It is a more sensible, more mature and more comprehensive view"

Charles Darwin.

7. ORIGIN OF STOCHASTICITY :

7.1. Introduction

So far there are two proposals to interpret quantum mechanics in terms of stochasticity :

(a) In Stochastic Electrodynamics (SED), [Pêna de La 1983,1995] it is proposed that there exists an all pervading electromagnetic zero point field over the whole universe and it might be the cause of stochastic nature of microparticles.

(b) The fluctuation of geometry is taken as an a priori notion and it seems to be the cause of stochasticity of microparticles.

It appears that the fluctuations of geometry and the hypothesis of random background field are completely independent and one has to choose either of the two as the source of stochasticity of microparticles.At the cosmological level, it is also assumed that due to density fluctuation of hadronic matter at early stage, a turbulent Gravito-Hydrodynamic [Nariai 1974, 1975] field might cause the fluctuation of the metric through Einstein equations.

At this point an enigma presents itself which agitated the great minds since the very inception of geometry itself. At the beginning of this century, Einstein raised this debate on a new level of understanding by formulating the general theory of relativity. In 1921, Einstein [Einstein 1921] delivered his famous lecture before Prussian Academy of Science on -"**Geometry and Experience**". Here, he used two kinds of geometry : one "purely axiomatic geometry" and the other "practical geometry".

In the "purely axiomatic geometry", the axioms are considered as free creation of human mind which should be taken in very formal sense i.e., as void of all contents of intuition or experience. On the otherhand, in "practical geometry", these are essentially based on induction from experience but not on logical inferences only. Thus geometry has become a part of natural science. Karl Menger [Menger 1979] in his article on "Geometry and Positivism" rightly pointed out :

"There should be a formalism for the refinement of geometry corresponding to the potential refinements of instruments and methods of observation ... A refinement of a geometry resolves some of its indeterminacies in the small ...".

In this chapter, we shall investigate the fundamental aspects of statistical geometry in the spirit of "Practical Geometry" as expounded by Einstein.

7.2. Random Zero Point Field

The idea that the random zero point field exists ubiquously throughout the universe attracted lots of attention since the original proposals of Planck and Nernst [Cetto A.M., 1993]. The possible connection of zero point field (ZPF) with the quantum properties of light has been studied since last few decades. On the otherhand, in SED, mainly the behaviour of microsystems was considered by various authors. Here it is assumed that the zero point radiation field is a real field. By considering the stochasticity, the ubiquity and the strength of the field, it is possible to show that the ZPF exerts a considerable change in the mechanical, thermodynamical and statistical description of matter. According to this picture, the electron or for the matter, any microscopic system that interacts electromagnetically, will have stochastic behavior and must be considered an essentially open system, which is permanently and stochastically interchanging energy and momentum with the vacuum field. The effect of stochastic forces, through nonlinearities, either of the forces or of the expressions that relate the dynamical variables, will affect also the systematic behavior of the system, and hence the laws of motion and the evolution will in general differ from the corresponding classical ones. Again as the fluctuations of ZPF are proportional to Planck's constant $\hbar$, the dispersions of the mechanical system induced by them will also be measured by $\hbar$ or its powers. Now it may be asked whether SED offer a physical explanation for the peculiar wave properties of matter?

Cetto et al. [Cetto, 1993] studied the possible action of this field on the particles and the conditions under which one can generate wave behaviour without loosing its corpuscular behaviour. Their starting point is to consider the dynamical effects produced by ZPF on a particle which responds to the modes of all frequencies up to a maximum ω_c, assumed to be fixed by the particle's effective structure. The particle is seen to vibrate under the effect of this field with wave number $k_c = \omega_c/c$. It is also suggested that the reletivistic calculations would have produced the Zitterbewegung. The coupling of the vibrating particle to the field modes of frequency ω_c, is shown to generate in the laboratory frame a modulated carrier wave with a wavelength given by de Broglie's formula. Moreover, the time-independent equation for this wave is shown to be stationary Schrödinger equation. The quantization rules for stationary states emerge from the condition of stationarity of the modulated wave.

Let us now discuss the mathematical description of the ZPF. Suppose the radiation field is taken in a cubic box of side L with perfectly conducting walls. For convenience let us take the Coulomb gauge. Then by decomposing the vector potential into plane waves with propagation vector E' and frequency $\omega = ck$, $k = |\vec{k}|$, we obtain

$$\vec{A} = \frac{1}{L^{\frac{3}{2}}} \sum_{k,\lambda} \hat{\epsilon}_{k\lambda}[c_{k\lambda} \exp i(\vec{k}.\vec{r} - \omega t) + c^*_{k\lambda} \exp -i(\vec{k}.\vec{r} - \omega t)] \tag{7.1}$$

where $\hat{\epsilon}_{k\lambda}$ are the polarization vectors, $\lambda = 1, 2$, $c^*_{k\lambda}$ complex conjugate of $c_{k\lambda}$. Again

$$\vec{k}.\hat{\epsilon}_{k\lambda} = 0, \quad \hat{\epsilon}_{k\lambda}.\hat{\epsilon}_{k\lambda'} = \delta_{\lambda\lambda'} \tag{7.2}$$

Now as $\hat{k} = \vec{k}/k$ and $\hat{\epsilon}_{k\lambda}$ form an orthonormal triplet, they satisfy the conditions

$$\sum_k \epsilon_{(k\lambda)_i}\, \epsilon_{(k\lambda)_j} = \delta_{ij} - \frac{k_i k_j}{k^2} \tag{7.3}$$

where

$$k_i = \frac{2\pi}{L} n_i; \quad n_i = 0, \pm 1, \pm 2, \cdots; i = 1, 2, 3$$

The amplitude $c_{k\lambda}$ are well defined number in classical field, operators for quantum field and random numbers in case of fluctuating ZPF. In the Coulomb gauge, the electric and magnetic fields can be given by

$$\vec{E}(\vec{r}, t) = -\frac{1}{c}\frac{\partial \vec{A}}{\partial t} \tag{7.4}$$

$$\vec{B}(\vec{r}, t) = \nabla \times \vec{A} \tag{7.5}$$

The above expression of vector potential $\vec{A}$ can be rewritten in a slightly different form as

$$\vec{A} = \sqrt{\frac{4\pi c^2}{L^2}} \sum_{n,\lambda} \hat{\epsilon}_{n,\lambda} \left[-q_{n\lambda} \cos \vec{k}_n.\vec{x} + \frac{\rho_{nk}}{\omega_n} \sin \vec{k}_n.\vec{x}\right] \tag{7.6}$$

where $\omega_n = ck_n$ Here, $q_{n\lambda}$ and $p_{n\lambda}$ are oscillatory functions of time that satisfy the equations

$$\begin{aligned} \ddot{q}_{n\lambda} + \omega_n q_{n\lambda} &= 0 & (7.7)\\ \ddot{p}_{n\lambda} + \omega^2 p_{n\lambda} &= 0 & (7.8)\\ \dot{q}_{n\lambda} &= p_{n\lambda} & (7.9)\\ \dot{p_{n\lambda}} &= -\omega_n^2 q_{n\lambda} & (7.10) \end{aligned}$$

Then the electric fields $\vec{E}$ and magnetic fields $\vec{B}$ can be decomposed as

$$\vec{E}(n,t) = \sqrt{\frac{4\pi}{L^3}} \sum_{n,\lambda} \hat{\epsilon}_{n\lambda} \left[p_{n\lambda} \cos \vec{k}_n.\vec{x} + \omega_n q_{n\lambda} \sin \vec{k}_n.\vec{x}\right] \tag{7.11}$$

$$\vec{B}(\vec{n};t) = \sqrt{\frac{4\pi c^2}{L^3}} \sum_{n,\lambda} (\vec{k}_n \times \hat{\epsilon}_{n,\lambda}) \left[q_{n\lambda} \sin \vec{k}_n.\vec{x} + \frac{p_{n\lambda}}{\omega_n} \cos \vec{k}_n.\vec{x}\right] \tag{7.12}$$

We shall now determine the statistical properties of $q_{n\lambda}$ and $p_{n\lambda}$. We can write

$$\begin{aligned} H_r &= \frac{1}{\sqrt{\pi}} \int (E^2 + B^2) d^3x \\ &= \frac{1}{2} \sum_{n,\lambda} (p_{n\lambda} + \omega_n^2 q_{n\lambda}^2) \\ &= \sum_{n,\lambda} H_{n\lambda} & (7.13) \end{aligned}$$

So, H_r can be thought of as sum of Hamiltonians of an infinite set of independent harmonic oscillators of unit mass with characteristic frequencies

$$H_{n\lambda} = 12 p_{n\lambda}{}^2 q_{n\lambda}$$

oscillators. Now in our case, $q's$ and $p's$ are stochastic variables and hence the value of $H_{n\lambda}$ is also a stochastic variable. We demand that the average energy of each one of these elementary oscillators of the field corresponds to the average energy of the QED vacuum field, namely $1/2\hbar\omega$. Taking average over the ensemble of all realizations of each elementary oscillators, we obtain

$$< H_{n\lambda} >= \frac{1}{2}\hbar\omega_n = \frac{1}{2} < p_{n\lambda} + \omega_n^2 q_{n\lambda}> \tag{7.14}$$

or equivalently

$$< p_{n\lambda}^2 >= \omega_n^2 < q_{n\lambda}^2 >= \frac{1}{2}\hbar\omega_n \tag{7.15}$$

Here, $<,>$ denotes the statistical average. In addition, there is a vacuum field for which $< q_{n\lambda} >= 0$; $< p_{n\lambda} >= 0$.

Assuming the Gaussian distribution of the amplitude q and p, we get the following statistical properties as

$$\begin{aligned} < q_{n\lambda}\, p_{n'\lambda'} > &= 0 \\ < q_{n\lambda}\, q_{n'\lambda'} > &= \frac{\hbar}{2\omega_n}\delta_{nn'}\,\delta_{\lambda\lambda'} \\ < p_{n\lambda}\, p_{n'\lambda'} > &= \frac{1}{2}\hbar\omega_n\delta_{nn'}\delta_{\lambda\lambda'} \end{aligned} \tag{7.16}$$

In a similar way, we can calculate the autocorrelation function of the electric field at two different points in space-time. We have

$$\begin{aligned} &< E_i(\vec{x},t)\; E_j(\vec{x}',t') >= \\ &\quad \sum_{n,n'} \sum_{\lambda,\lambda'} f(\omega_n) f(\omega_{n'})\; \epsilon_{(n\lambda)_i}\, \epsilon_{(n'\lambda')_j} \\ &\quad < a_{n\lambda}\, a_{n'\lambda'}^*\; \exp[i(\vec{k}_n.\vec{x} - \omega_n t) - i(\vec{k}_{n'}.\vec{x}' - \omega_{n'} t')] \\ &+ \; a_{n\lambda}^*\, a_{n'\lambda'} \exp[-i(\vec{k}_n.\vec{x} - \omega_n t) + i(\vec{k}_{n'}.\vec{x}' - \omega_{n'} t'] > \end{aligned} \tag{7.17}$$

where

$$\begin{aligned} p_{n\lambda} &= \frac{1}{2}\sqrt{\hbar\omega_n}(a_{n\lambda}^*\; \exp[i\omega t] + a_{n\lambda}\exp[-i\omega t]) \\ q_{n\lambda} &= \frac{\sqrt{\hbar\omega_n}}{2i\omega_n}(a_{n\lambda}^*\exp[i\omega_n t] - a_{n\lambda}\exp[-i\omega_n t]) \end{aligned} \tag{7.18}$$

After some straightforward calculations

$$< E_i(\vec{x},t)E_j(\vec{x}',t') > \;=\; < B_i(\vec{x},t)B_j(\vec{x}',t')) >$$

$$= \frac{2\pi\hbar}{L^3}\sum_n \Omega_n(\delta_{ij} - \frac{k_{ni}k_{nj}}{k_n^2}) \times \cos[\vec{k}_n.(\vec{x} - \vec{x}') - \omega_n(t - t')] \quad (7.19)$$

Similarly we can also calculate the two point correlation for magnetic field. In the limit $L \to \infty$, the above two point correlation function can be written as

$$< E_i(\vec{x},t)E_j(\vec{x}',t') >= \frac{\hbar}{4\pi^2}\int(\delta_{ij} - \frac{k_ik_j}{k^2})\omega\cos[\vec{k}.(\vec{x} - \vec{x}') - \omega(t - t')]d^3k \quad (7.20)$$

Taking

$$\oint(\delta_{ij} - \frac{k_ik_j}{k^2})d\Omega_k = \frac{8\pi}{3}\delta_{ij}$$

the two point correlation function further reduces to

$$< E_i(\vec{x}.t)E_j(\vec{x}',t') >= \frac{2\hbar}{3\pi c^3}\delta_{ij}\int_0^\omega \omega^3\cos(t - t' - \frac{|\vec{k} - \vec{k}'|}{c})\omega d\omega \quad (7.21)$$

It has been recently proposed [Roy 1992] that the random zero point field produces the probabilistic aspect in geometry. Here we shall consider Rander Space [Rander 1941] where the metric tensor becomes the function of $A_\mu(x)A_\nu(x)$, $A_\mu(x)$ is the electromagnetic potential. If we take $A_\mu(x)$ as random potential, as considered in SED, with certain statistical properties, then there will be a non-zero contribution to the metric tensor by averaging over the ensemble of fields $A_\mu(x)$. Let

$$a(x,y) = (a_{ij}(x)Y^iY^j)^{\frac{1}{2}}$$

$$b(x,y) = b_i(x)Y^i$$

where, a_{ij} is a Riemannian metric tensor and $b_i(x)$ is a covariant vector field. Then we can construct the metric function as

$$F(x,y) = a(x,y) + b(x,y) \quad (7.22)$$

which is known as Rander's metric function. Here, $F(x,y)$ is a non-negative scalar function defined over a real N-dimensional differentiable manifold M. The two sets of arrangements of $F(x,y)$ are namely the points x^i and contravariant vectors y^i, tangent to M at x^i or symbolically, $x^i \in M$ and

$y^i \in T_x$. The associated geodesic equation will be written in terms of Lagrangian function

$$F(x, \dot{x}) = a(x, \dot{x}) + b(x, \dot{x}) \tag{7.23}$$

where, $\dot{x} = \frac{dx}{dq}$, q a parameter along a curve.

The metric tensor related to the above metric function can be written as

$$g_{ij} = F(a_{ij} - u_i u_j)/a + p_i p_j \tag{7.24}$$

and similarly for g^{ij}. Here,

$$b_i = a^{ij} b_j; \quad u_i = \frac{a_i}{y} = a_{ij}\frac{y^j}{a} \tag{7.25}$$

$$p_i = \frac{\partial F}{\partial y} = u_i + b_i; \quad p^i = a^{ij} p_j \tag{7.26}$$

If we restrict ourselves to the four-dimensional case $N - 4$, identifying $a_{ij}(x)$ with the (pseudo) Riemannian metric tensor, taking $b_i = \frac{eA_i(x)}{m_0 c^2}$, the metric function $F(x, \dot{x})$ becomes

$$F(x, \dot{x}) = (a_{ij}(x)\dot{x}^i \dot{x}^j)^2 + \frac{e}{m_0 c^2} A_i(x)\dot{x}^i \tag{7.27}$$

This expression is, up to a constant factor, the Lagrangian function of a test electric charge in the electromagnetic and gravitational fields described by the vector potential and the Riemannian metric tensor $a_{ij}(x)$.

Now, as $A_i(x)$ has been taken as random function (due to the existence of random zero point field), the metric tensor g_{ij} becomes random function.

$$\begin{aligned} g_{ij} &= \frac{FA_{ij}}{a} - Fu_i u_j a + u_i u_j + u_i \frac{e}{m_0 c^2} A_j(x) \\ &+ \frac{e}{m_0 c^2} A_i(x) u_j + \frac{e}{m_0{}^2 c^4} A_i(x) A_j(x) \end{aligned} \tag{7.28}$$

In Stochastic Electrodynamics the statistical properties of the potentials [Pêna 1995] are taken as $< A_i(x) >= 0$. Now, for Coulomb gauge, $< \vec{A} >= 0$ and using the statistical properties as described in (7.21) we get

$$< A_i(x_1) A_j(x_2) >= \frac{2\hbar}{3\pi c} f(x_1 - x_2)\delta_{ij}$$

where, $i = 1, 2, 3$; $j = 1, 2, 3$ and

$$f(x_1 - x_2) = \int_0^\infty \int_0^\infty \int_0^\infty \omega^3 d\omega \cos\omega[(t_1 - t_2) - (|x_1| - |x_2|)] d\omega dt_1 dt_2 \quad (7.29)$$

Taking the statistical average of g_{ij} over the ensemble of $A_i(x)$,

$$< g_{ij} >= \bar{g}_{ij} = \bar{a}_{ij}(x) + \frac{e^2}{m_0{}^2 c^4} f(x_1 - x_2) \frac{2\hbar}{3\pi c} k_{ij} \quad (7.30)$$

where, $k_{ij} = k_{22} = k_{33} = 1$; $k_{44} = 0$, in a similar manner, for pseudo-Euclidean space, the resulting metric tensor is

$$\bar{K}_{ij} = \eta_{ij} + \frac{e^2}{m_0{}^2 c^4} f(x_1 - x_2) \frac{2\hbar}{3\pi c} k_{ij} \quad (7.31)$$

However, the resulting metric tensor $\bar{g}_{ij}$ or $\bar{K}_{ij}$ will not satisfy the property of the metric itself because of the presence of the divergent integral in the second term of right hand side. In otherwords, the space-time will loose its metrical property for high frequencies or at short distances. At this stage, the topological fluctuation of the space time might be very large and the space is not metrizable. So, to retain the metrical property of the space-time, there should be a cut-off in the energy spectrum of the zero point field. Up till now, it has been merely a belief that some kind of cut-off rule is operating in QED at short distances. But here, the metrizability of the space-time ensures the cut-off in the frequency spectrum of the zero point field.

If m_0 be taken as the mass of the electron, then,

$$\bar{K}_{ij} = \eta_{ij} + 1.37 \times 10^{-2} \times k_f k_{ij} \quad (7.32)$$

where, k_f is the momentum cut -off corresponding to the cut off frequency so as to retain the metrizability of the space-time. Also, $k_f \sim \frac{1}{\ell_f}$, ℓ_f being the cut off length which corresponds to the length scale of fluctuation of metric. Comparing this metric with the Lorentz non-invariant part of the metric considered by Nielson and Picek [Nielson 1983], we get

$$\ell_f \sim 10^{-27}\text{cm}$$

which is much greater than Planck length.

(A) Implications of Cut-off Length :

Performing a measurement analysis, it has been shown by Jordan and Fock [Jordan 1930] that using single electron or proton as test bodies to measure the strength of electromagnetic field, it puts on limitations on the measurability of these fields. In the case of the electric field, read

$$\Delta E_x \Delta x \Delta t \geq \frac{\hbar}{e} \tag{7.33}$$

and for the magnetic field

$$\Delta H \Delta y a \geq \frac{\hbar c}{a} \tag{7.34}$$

$\Delta E_x \cdots, \Delta H_x \cdots$ being uncertainties of the electromagnetic field strength resulting from the momentum uncertainty given by the Heisenberg's relation $\Delta p_x \Delta x \geq \hbar$, while δt is the time interval between two momentum measurements necessary for the determination of $E_x \cdots$ and a, the radius of the path of the test body moving in the magnetic field. Calling δt and a as Δt and Δx, it can be shown that a single field component i.e., E_x can not be defined sufficiently exactly at an exactly defined space-time point x, y, z, t. The type of uncertainty relations defined above are difficult to interpret because they connect quantum theoretically defined uncertainties i.e., in the 1st relation (eqn. 7.33) ΔE_x and Δx with Δt and in the 2nd one (eqn 7.34) ΔH_x andΔy with a respectively. However, in the first one, Δt is not on the same theoretical footing with ΔE_x and Δx; in the 2nd case there is interpretative difference between ΔH_x and Δy, on the one hand, and a, on the otherhand. While the first quantities refer to the system to be measured (object system), the latter are quantities characterizing the measurement device. Therefore, the above relations can not be interpreted like usual Heisenberg uncertainty relations

$$\Delta p_x \Delta x \geq \hbar; \quad \Delta p_y \Delta y \geq \hbar; \quad \Delta p_z \Delta z \geq \hbar$$

but rather like of the type

$$\Delta E \Delta t \geq \hbar \tag{7.35}$$

In QED, this interpretative problem, however, is an apparent problem. Bohr and Rosenfeld [Bohr 1933] analysed the above relations and showed that the measurement analysis providing first type relations in electrodynamics does not concern optimal measurement. Thus these relations do not have a fundamental meaning, that means, do not have a direct correspondence in the quantum formalism.

In the quantized general relativity, the situation changes drastically. By reanalyzing the optimal field measurement, it leads to Jordan-Fock type relation, namely to

$$\Delta R . L_0{}^3 \geq \left(\frac{\hbar G}{c^3}\right)^{\frac{1}{2}} \text{ etc.} \tag{7.36}$$

Here, ΔR denotes a quantum measurement inaccuracy of the Ricci scalar

$$R = g^{\mu\nu} R_{\mu\nu}$$

and L_0 is the extent of the measurement body. The fact that Jordan-Fock type uncertainty relation, in contrast to electrodynamics, are satisfied for optimal measurements in quantized general relativity signals the great difference between quantum electrodynamics and quantum general relativity and also the difficulties one meets within the latter case. But just because of this circumstances, it is useful to study the above relation. Instead of studying the relation as such, we have considered the generalized version i.e.,

$$1 > \ell_0 \equiv \frac{1}{L_0{}^3} (\ell_p L)^2; \quad \ell_p \equiv \left(\frac{\hbar G}{c^3}\right)^{\frac{1}{2}} \tag{7.37}$$

This generalization results by regarding not only the weak principle of equivalence but also the strong one. This generalized relation reflects the fact that due to the equivalence principle, all matter moves in an effective Riemannian space of metric

$$g_{ij} = \eta_{ij} + h_{ij} \tag{7.38}$$

so that no (either a flat or a given curved) background is observable. Considering covariant background quantization of general relativity one starts with the splitting of the metric into a background field η_{ij} and a quantum field h_{ij}. But this splitting is lost at distances ℓ_0 given by the above relation.

(B) A Cut-off of High Frequency and Strong Field Effects in Quantum General Relativity:

(a) As far as gravitational quantum effects like gravitational Compton effect, Bremsstrahlung, Pair creation and Lamb shift are concerned as a consequence of length ℓ_0, the high frequency part of these effects can not be measured. In compton effect,

$$\frac{1}{\nu'} = \frac{1}{\nu} + \frac{\hbar}{mc}(1 - \cos\theta) \tag{7.39}$$

ν and ν' being the frequencies of the gravitational wave before and after scattering respectively, $\frac{\hbar}{mc}$ is the wavelength of the scattering particles and θ is the scattering angle. The occurence of this cut-off length to restrict the validity of the above formula to regions satisfying the relation

$$\lambda\Lambda >> \ell_0{}^2$$

where, $\Lambda \equiv \frac{\hbar}{mc}$. For an optimal measurement,

$$\ell_0 = (\ell_p L)^{\frac{1}{2}}$$

one sees that one can get a cut-off at distances much greater than Planck length ℓ_p.

(b) The same is true for Gravitational Vacuum Polarisation Effects. By virtue of our generalized relation, such effects are cut-off at distances

$$\lambda >> (\frac{\Lambda'}{d_g})^{\frac{1}{2}} \ell_p = \Lambda' >> \ell_p$$

$\Lambda' = \frac{\hbar}{mc}$ is the Compton wave length of virtual particles, $d_g = \frac{GM}{c^2}$ be their gravitational radii and λ, the wavelength of the quantum fluctuations h_{ij}. This shows that there is no abrupt cut-off at Planck length but the gradual cut-off, where Planck's order of magnitude denote only the last, absolute limit.

(c) Assuming that the metric operator $\hat{g}_{ij}$ is again splitted into a classical background field γ_{ij} and quantum perturbations $\hat{h}_{ij}$, then in the thermodynamic equilibrium, the states $|\Phi>$ of the $\hat{h}_{ij}$ field, satisfying the equation [Borzeszkowski 1988]

$$H|\Phi>= \Phi|\Phi> \tag{7.40}$$

must be occupied according to the formula

$$n = \frac{1}{\exp(\frac{\Phi}{kT}) - 1} \tag{7.41}$$

Accordingly, the spectral density of the field must be given by Planck formula

$$\rho = \frac{8\pi h\nu^3}{c^3} . \frac{1}{\exp(\frac{\Phi}{kT}) - 1} \tag{7.42}$$

Considering however, this formula is the high frequency region near distance ℓ_0; it can be shown that the distribution is not stable there.

7.3. Stochastic Strings :

Recently, new interest arises to study the space-time structure at short distances due to the developments of the theory of strings and superstrings [Witten 1996]. In string theory, the one-dimensional trajectory of a particle in space-time is replaced by a two-dimensional orbit of a string. Such a string can be of any size, they may be as tiny as 10^{-32}cm in diameter. This is larger than Planck length but comparable with the length scale of fluctuation of the metric as estimated by several authors. In string theory, one introduces a new fundamental constant

$$\alpha' = (10^{-32}\mathrm{cm})^2$$

controlling the tension of the string. It is widely believed that new physics may emerge when string theory is expressed at such a small length scale. Perhaps space-time itself will loose its meaming at very short distances, In fact, if we wish to understand the underlying principle and symmetries of string theory we must explore its behavior at its charachteristic mass scale namely the Planck scale. But one of the main problems to explore the physics at the Planck scale is the absence of experiments that directly probe this region. One can explore the mathematical structure of the theory with the hope that this will reveal new structures and methods. The Gadenken experiment may also be designed to probe such a small scale structure. However, explorations of the short distance structure of string theory are not easy due to the fact that we can not couple exernal currents to strings in a consistent way which would allow for an operational way of defining local observables. This is inevitable in a theory of everything, since if we could introduce external probes into the theory it could be modified and would no longer be unique. However, the theory is so symmetric that no modification appears possible. To explore the structure of strings we can only use as probes the object which are contained within the theory.

The recent advances in formulating covariant string theory suggest that we can study short-distances behavior by constructing the two point function or propagator of the relevant field operators. The analog in string theory would be

$$< \Omega|\Psi[x(\sigma')]\Psi[x(\sigma)|\Omega > \tag{7.43}$$

$\Psi[x(\sigma)]$ is a string field. Using either the older light cone formulations of string field theory or the newer covariant approaches it is easy to show that the propagator has unphysical (space.like) singularities whenever

$$\int (\frac{d\sigma}{\pi})[x(\sigma') - x(\sigma)]^2 \leq \frac{c}{{m_{\mathrm{Planck}}}^2} \tag{7.44}$$

These singularities are, of course, not present in the on shell amplitudes. Their presence in the off-shell extrapolations indicates that this field might not be very useful in probing the short distance structure. So we are limited to the study of on-shell scattering amplitudes. Gross et al. [Gross, 1988] studied the behavior of the string scattering amplitudes to all orders in perturbation theory, with the aim of exploring the short-distance structure. It is shown that the sum over all Riemann surfaces is dominated by saddle point. So saddle point techniques can be used to probe the structure of string theory at Planck scale. An interesting space-time structure emerges at high energy limit. In space-time we found a surprising picture of how strings, which are extended objects, get close. Stochastic geometry plays an important role in representing gauge theories by random surfaces and strings [Polyakov 1981], [Gomez 1982].

It is important to note that in this monograph we have considered the space-time as a set of hazy lumps instead of point set. This is known a cloud space or C-space. The hazy lumps are considered as the seat of elementary particles. The essential feature of this theory would be that lumps would not be point set. They would rather be in mutual probabilistic relations of overlapping and apartness, from which a metric would have to be developed.

In C-space, these hazy lumps are considered as elements of space instead of "ordinary" points. Now if one identify these hazy lumps with the random strings, then a probabilistic structure can be built up at the microlevel. Let us consider a bosonic string, the motion of which can be described by a two dimensional surface known as the string world sheet. This string world sheet is embedded in a d-dimensional space-time (for example $d = 4$). Now this manifold M is parametrized by the coordinates

$$\sigma = \sigma^\alpha \equiv (\sigma^1 = \sigma,\ \sigma^2 = \tau)$$

and is equipped with a metric tensor $g_{\mu\nu}$ while the space-time has coordinates $x^\mu(\tau)$ and metric $G_{\mu\nu}$. Here we need two parameters instead of single parameter τ for world lines of particles. In fact, the world lines are replaced by world-tubes of strings. Now as the world tube is two-dimensional, one has not only just a proper time τ along the trajectory but a proper position σ also. Then the motion of the string can be described by the functions $y^\mu(\sigma^\alpha)$.

In case of random strings, $y^\mu(\sigma^\alpha = \sigma)$ are considered [Namsrai 1996] as random variables. The probability distribution $P(y)$ can be written as

$$P[y] = \frac{1}{N}\exp\left\{-\frac{1}{2}\int_{M_1}\int_{M_2} d^2\sigma_1 d^2\sigma_2 \sqrt{g_1}\sqrt{g_2}\, y^\mu(\sigma_1) D^{-1}_{\mu\nu}(\sigma_1 - \sigma_2) y^\nu(\sigma_2)\right\} \tag{7.45}$$

N is the normalization constant and $D^{-1}_{\mu\nu}$ is the inverse of two-point correlation, given by

$$< y^{\mu}(\sigma_1)y^{\nu}(\sigma_2) >= D^{\mu\nu}(\sigma_1 - \sigma_2) \tag{7.46}$$

The main assumption is that the space-time fluctuates due to the presence of random string field. So the topological structure gives rise to changes in physical quantities at short-distances. In order to introduce stochastic fluctuation in the metric induced by the random strings, we now define confomal-like transformation leading to the passage from the usual local inertial system of reference ξ^{a} with the Minkowski metric $\eta_{\alpha\beta}$ to the averaged string coordinate x^{μ} :

$$y^{\mu}(\sigma^1, \sigma^2) = (\frac{1}{\sqrt{g}}R)\delta(\sigma_1)x^{\mu}(\sigma_2) + \zeta^{\mu}(\sigma_1, \sigma_2) \tag{7.47}$$

with an induced stochastic metric $G_{\mu\nu}(x, y)$ and ζ^{μ} as random variables. Let us consider the transformation

$$\xi^{\mu} = x^{\mu} \exp\left\{\frac{1}{2(\pi a^1)^{1/2}} \int_{\mu} d^2\sigma\sqrt{g}RU_{\mu}(\sigma)y^{\mu}(\sigma)\right\} \tag{7.48}$$

R is the Ricci curvature of the manifold M and $U_{\mu}(\sigma)$ is a unit vector with,

$$-\eta^{\mu\nu}U_{\mu}(\tau)U_{\nu}(\tau) = 1 \tag{7.49}$$

The rectilinear trajectory of a particle is given by

$$\frac{d^2\xi^{\alpha}}{d\tau^2} = 0, \quad d\tau^2 = -\eta_{\alpha\beta}d\xi^{\alpha}d\xi^{\beta} \tag{7.50}$$

In the system of reference ξ^{a} , it can be written as

$$\frac{d^2x^{\lambda}}{d\tau^2} + \Gamma^{\lambda}_{\mu\nu}\frac{dx^{\mu}}{d\tau}\frac{dx^{\nu}}{d\tau} = 0 \tag{7.51}$$

where $\Gamma^{\lambda}_{\mu\nu}$ is the affine connection-like quantity defined as

$$\Gamma^{\lambda}_{\mu\nu} = \frac{\partial x^{\lambda}}{\partial \xi^{\alpha}}\frac{\partial^2\xi^{\alpha}}{\partial x^{\mu}\partial x^{\nu}} \tag{7.52}$$

Now the proper-time can be expressed as

$$d\tau^2 = -G_{\mu\nu}(x, y)dx^{\mu}dx^{\nu} \tag{7.53}$$

where $G_{\mu\nu}(x,y)$ is the stochastic metric.

$$G_{\mu\nu}(x,y) = \eta_{\alpha\beta}\frac{\partial\xi^{\alpha}}{\partial x^{\mu}}\frac{\partial x^{\beta}}{\partial x^{\nu}} \tag{7.54}$$

It can be shown that [Namsrai 1996]

$$G^{\nu\alpha}(x,y)G_{\beta\nu}(x,y) = \delta^{\alpha}_{\beta} \tag{7.55}$$

But it is necessary to specify the averaging procedure over random variables $y^{\mu}(\sigma)$ with probability distribution $P(y)$.

For white noise, the two point correlation function can be written as

$$D^{\mu\nu}(\sigma_1 - \sigma_2) = -\eta^{\mu\nu}\frac{\lambda^2}{\sqrt{g}R}\delta^2(\sigma_1 - \sigma_2) \tag{7.56}$$

$\eta^{\mu\nu}$ is Minkowski metric tensor as stated above, λ is a constant and has dimension of length.

There may be two possibilities for the following values of λ as :

- (i) $\lambda^2 \simeq G^2, G$ is the Newtonian constant.
- (ii) $\lambda^2 \approx (\sqrt{\alpha'})^2$, $\sqrt{\alpha'}$ is the inverse string tension (a size of the string).

In case (i), the length scale of fluctuation of the string coordinates is of the order of Planck scale. On the otherhand, in the case (ii), this scale of fluctuation is of the order the size of the string.

Now the physical space-time metric at large distances can be obtained as follows:

Using Feynman's rules

$$\langle \exp\int d^4x K(x)\Psi(x)\rangle = \exp(\frac{1}{2}\int\int d^4x d^4y K(x)\Delta(x-y)K(y)) \tag{7.57}$$

we have

$$\langle G_{\mu\nu}(x,y)\rangle_y = [\eta_{\mu\nu} + \epsilon_{\mu\nu}(x) + \frac{1}{4}\epsilon^{\rho}_{\mu}(x)\epsilon_{\nu\rho}(x)]\exp(\frac{\lambda^2}{2\pi\alpha'}\int_{\mu} d^2\sigma\sqrt{g}R) \tag{7.58}$$

where

$$\text{Euler}(\mu) = \frac{1}{4}\int d^2\sigma\sqrt{g}R$$

Thus the physical space-time metric at large distances can be written as

$$G_{\mu\nu}(x) = \sum_N \langle\langle G_{\mu\nu}(x,y)\rangle\rangle_y \tag{7.59}$$

$$\begin{aligned} G_{\mu\nu}(x) &= \sum_N < G_{\mu\nu}(x,y) >_y \\ &= [\eta_{\mu\nu} + \epsilon_{\mu\nu}(x) + \frac{1}{4}\epsilon^{\rho}_{\mu}(x) \in_{\rho\nu} (x)] \times I \end{aligned} \tag{7.60}$$

where

$$I = \frac{e^{4\Delta}}{1 - e^{-4\Delta}}, \quad \Delta = \frac{\lambda^2}{\alpha'}$$

This multiplier I appears due to the topological structure of space-time at small distances. It may be mentioned that by using general covariant method of the description of gravitational phenomena in space-time with stochastic metric in the weak-field limit with metric $G_{\mu\nu}(x,y)$, it is possible to construct the theory of induced gravity caused by random strings. It is also observed that due to the fluctuation of the space-time metric associated to random strings, the quarks can be confined inside the region characterized by the string tension parameter α'.

7.4. New Correspondence Principle :

Recently, Witten [Witten 1996] considered an interesting situation regarding the duality principle and the string tension parameter α'. He described it as :

"There is a smallest circle in string theory; a circle of radius is equivalent to a circle of radius α'/R. By this we mean most simply the following : Imagine that the universe as a whole is not infinite in spatial extent, but one of the three space dimensions is wrapped in a circle, making it a periodic variable with period $2\pi R$. Then there is a smallest possible value of R. When R is large, things will look normal, but if one tries to shrink things down until the period is less than $2\pi\sqrt{\alpha'}$, space will re-expand in another "direction" peculiar to string theory and one will not really succeed in creating a circle with a radius of less than $\sqrt{\alpha'}$.... There is a duality symmetry - generalizing the duality of Ising model - that exchanges the two spectra, exchanging also R with α'/R".

It seems that one cannot compress a circle below a certain length. The Heisenberg microscope has been reanalyzed and it is claimed that Heisenberg microscope does not work in the string theory if the energy is too large. It is speculated that as if the uncertainty principle has two terms instead of one i.e.,

$$\Delta x \geq \frac{\hbar}{\Delta p} + \alpha' \frac{\Delta p}{\hbar}$$

Here, the first term is familiar quantum uncertainty and the second term reflects a uncertainty or fuzziness due to string theory. Now with the two terms together, there is an absolute minimum uncertainty in length of order $\sqrt{\alpha'} = 10^{-32}\text{cm}$ in any experiment. If the string theory is correct, then there will be two constants in real world :

- Planck constant $\hbar$ ($\hbar \neq 0$) in quantum mechanics
- Non-zero α' in string theory.

Bohr introduced the idea of correspondence principle in studying the transition from classical to quantum domain. He proposed that in the limit $\hbar \to 0$, we can explain the transition from quantum to classical domain. Here, we are proposing a new correspondence principle in the limit $\alpha' \to 0$. Again, the space-time might be fluctuating due to the existence of quantum fluctuations at very short distance [Percival 1995]. So, the new correspondence principle may help us to understand even the nature of quantum fluctuations. But the element-set relation should be rigorously studied in the limit $\alpha' \to 0$ before discussing the detail consequences of the newly proposed correspondence relation.

Here we have considered the hazy lumps as the size of the string and hence a fluctuating metric. So the limit $\alpha' \to 0$ implies a kind of transition from the space as a set of hazy lumps to a set of points and we arrive at the continuum. In this way, it may be possible to resolve the controversy between physical continuum and mathematical limiting procedure. Now with $\alpha' \neq 0$, if we consider the limit $\hbar \to 0$ we can construct a probabilistic metric space which will be classical in nature. This is similar to the conclusion drawn by Belinskii et al. [Belinskii 1982] in the context of Einstein equations for classical gravitational field and its solutions near time singularity region which will be discussed in chapter 8.

7.5 Stochastic Background Force and Origin of Quantization

Recently Calogero[Calogero 1997] published a thought provoking paper on the physical basis for the background field hypothesis as proposed by Nelson [Nelson 1985]. Planck constant has been shown to be related to the gravitational constant and the radius of the universe. The only cosmological parameter enters is the radius of the universe. So Planck constant is no longer considered as the fundamental constant in nature. This gives rise to new insight on the origin of the quantization itself via the stochastic quantization procedure. This analysis is mainly based on dimensional considerations. The main analysis of Calogero is based on the following elementary considerations.

(1) The gravitational interaction is considered as universal in the sense that it has an equal kinematical effect on every element of matter (particle).

(2) The classical mechanics with standard gravitational force is assumed to be the underlying theory describing the physical world.

(3) The granularity hypothesis i.e. the universe has a granular structure. It means that main components are taken as particles.

(4) Classical mechanics entails that the motions of many body systems are chaotic, hence, the motion of every particle in the Universe must have a chaotic component at least to the extent it is governed by classical mechanics.

(5) As a consequence of these facts, every particle experiences locally essentially the same stochastic gravitational acceleration, due to the rest of the universe; in addition of course to non stochastic effects of gravitation,as well as other interactions (electromagnetic, weak and strong). So one can think of a universal stochastic background interaction where the spatial coherence of the tiny tremor of matter can be maintained by long range gravitational force. The universal stochastic background may have its origin as gravitation but the main problem is to construct a quantity having the dimension of action (which is Planck constant in stochastic quantization) by taking gravitational constant G and speed of light c. Using dimensional considerations Calogero[Calogero 1997] made a semi-quantitative theory based on the above elementary considerations. Let us consider the following relevant quantities : Gravitational Constant G :

$$G = \frac{\text{length}}{\text{mass time}^2}, \quad M : \text{mass of the Universe}$$

$$D : \text{Its average density}; \quad R : \text{Radius of the Universe}$$

$$R = b^2 (\frac{M}{D}^{\frac{1}{3}} \qquad b = \frac{3}{4\pi}^{\frac{1}{6}} = 0.7876$$

$$T = b^3 (GD)^{-\frac{1}{2}} \quad V = \frac{R}{T} \quad \text{acceleration} : a = 3b^{-4} GMD^{\frac{2}{3}}$$

$$\text{energy} : E = b^{-2} GM^{\frac{5}{3}} D^{\frac{1}{3}} \quad \text{action} : A = b^{-9} G^{\frac{1}{2}} D^{\frac{3}{2}} R^5$$

Where

$$a = G \int_0^R \frac{D 4\pi r^2 dr}{r^2}$$

is the gravitational acceleration experienced by any prticle at the center of a spherical universe of radius R and density$D$. Now it is important to estimate the charcteristic time τ of the stochastic motion. The chaotic character of this motion is due to the granularity of the universe. Hence, it is plausible to expect that the characteristic frequency of this motion be larger, the larger is the number N of particles in the universe. It is assumed to be proportional to the square root of N. Then,

$$\tau \cong n^{-\frac{1}{2}}T = b^3 N^{-\frac{1}{2}} D^{-\frac{1}{2}} \tag{7.61}$$

with $b = 0.7876$.

Now to estimate the quantum of action h associated within stochastic motion of each particle we multiply by τ the gravitational energy ϵ of each particle as

$$\epsilon^2 \cong \frac{E}{N} = N^{-1}GM^2R^{-1} = b^{-2}n^{-1}GM^{\frac{5}{3}}d^{\frac{1}{3}} \tag{7.62}$$

Here, it is assumed that total gravitational energy E of the Universe is equally partitioned between N particle of the universe. we can write the action h as

$$h = \epsilon\tau \cong N^{-\frac{3}{2}}A = bN^{-\frac{3}{2}}G^{\frac{1}{2}}M^{\frac{5}{3}}D^{-\frac{1}{6}} \tag{7.63}$$

This can be rewritten as

$$h \cong G^{\frac{1}{2}}m^{\frac{3}{2}}R^{\frac{1}{2}} \tag{7.64}$$

by putting $N = \frac{M}{m}$.

Here, we have replaced the number of particles N by the ratio of the mass of universe (M) and the nucleon or of the Hydrozen atom mass (m). In this case, only cosmological quantity enters is the radius of the universe. It should be noted that this formula was long back considered by Weinberg [Weinberg 1972]. But its significance was not propaerly understood at least in the perspective of stochastic quantization. We can now make a numerical estimate for action h taking the following values for m, G, R, etc.,

$$\begin{aligned}
m &= 1.67 \times 10^{-24} g \\
G &= 6.6 \times 10^{-8} g^{-1}\mathrm{cm}^3 s^{-2} \\
R &\cong 10^{\rho}\mathrm{cm}, \quad \rho = 28 \pm 2 \\
D &\cong 10^{-\delta}\mathrm{gcm}^{-3}, \quad \delta = 30 \pm 2 \\
M &\cong 4 \times 10^{\mu} g, \quad \mu = 54 \pm 8 \\
N &\cong \frac{M}{m} \cong 2.5 \times 10^{\nu}, \quad \nu = 78 \pm 8
\end{aligned}$$

Then

$$h \cong 6 \times 10^{-p} g\mathrm{cm}^2 s^{-1}; \qquad p = 26 \pm 1$$

This value of Planck constant is remarkably close to the actual values ($h \approx 6 \times 10^{-27} g\mathrm{cm}^2 s^{-1}$). We must emphasize that the estimated value of h is very close to the observed value inspite of the grossly overestimated uncertainty associated with cosmological quantities. In a similar manner we can estimate the length scale associated to the random motion as

$$l \cong 10^{-\lambda}\mathrm{cm}, \qquad \lambda = 50 \pm 6$$

which is so small to escape the observational limit.

We have already mentioned that the gravitational interaction helps us to build up a coherent model of the universeal tremor as consisting merely of a uniform " streching / shrinking /breathing mode" of the universe, such that, if its mass M remains constant, its matter density D undergoes a tiny uniform overall fluctuation according to the formula

$$D = D_0 \left[1 + N^{-1} f(\frac{t}{\tau)}\right] \tag{7.65}$$

and this density fluctuation may lead to the fluctuation of the metric.

In chapter **2**, we have discussed the role of fluctuation of matter density in dephasing the interference in Neutron Interferometer [Rauch, 1995]. It would be interesting to consider the effect of this fluctuation in decoherence since this fluctuation is universal and affect any particle in the universe. Moreover, if this density fluctuation is considered as the cause of fluctuation of the metric, then this approach may help us to solve the problem of metrizability as considered in section **7.2** without introducing an adhoc cut-off length.

The most attractive feature of this approach is the "granularity effect" which is identified as the cause of stochasticity. This kind of stochasticity may also be considered as the cause of fluctuation of metric. However, as this analysis is based mainly on semi-quanlitative and sometimes semi-quantitative arguments, it needs more efforts to have a rigorous physical basis of stochastic quantization procedure.

POSSIBLE IMPLICATIONS

"Perhaps quantum fluctuations are just as real as thermal fluctuations and arise from certain interactions ..."

E. Nelson : Quantum Fluctuations.

8. Possible Implications :

8.1 Prologue :

Throughout this book we have repeatedly emphasized the role of fluctuations in explaining the statistical behavior of microparticles as well as its role at microscopic and megascopic scales. Prigogine et al. [Prigogine 1984] discussed extensively the role of fluctuation and irreversibility in his thought provoking monograph **"Order out of chaos"** - man's discourse with nature. He mainly concentrated his discussion in the context of non-equilibrium processes. At the end of nineteenth century, most scientists were baffled by the fact that how the formulation of thermodynamics can be reconciled with dynamics. It was Boltzmann who thought that new concepts had to be developed to extend the physics of trajectories (in dynamics) to cover the situation described by thermodynamics. Boltzmann sought this conceptual innovation in the theory of probability. Since then, the role of fluctuations becomes the part of a great debate not only in the microscopic scale but also in macroscopic phenomena.

Ehrenfest proposed a model popularly known as **"Urn model"** to clarify Boltzmann association of probability and entropy. But Ehrenfest model is nothing but a simple example of a "Markov Process". The basic ingredient of Markov process is the assumption that the transition probabilities are independent of the previous history of the system. In Ehrenfest model, it is the probability distributions, not the individual events, evolves irreversibly.

Therefore, on the level of distribution functions, Markov chains lead to a one wayness in time.

The amazing aspect of this story is that the same idea of Markov chains can be used to explain the temporal evolution of wave function in quantum mechanics where the wave function evolves reversibly. So the fluctuation plays a different role in quantum mechanics than thermodynamics. This is considered as quantum fluctuation which is obviously non-dissipative in nature. In this chapter, we shall discuss the role of fluctuation of metric in general and the non-dissipative nature of fluctuations in case of quantum particles.

8.2 Quantum Fluctuation :

At the end of his classic book on "Quantum Fluctuations", Nelson [Nelson 1985] posed a set of open problems beginning with

"To find a classical Lagrangian, of system + background field oscillators + interaction, that with reasonable initial probability measures and in the limit as the cut offs on the background field are removed, produces a conservative diffusion in the system or to show that this is impossible ...".

Since the very beginning of quantum mechanics, the similarities between the classical equations describing Brownian motion of a particle (i.e., a kind of diffusion equation) and Schrödinger description for microparticles attracted the great minds time and again. It is Nelson [Nelson 1967] who gave a rigorous mathematical foundation of stochastic model of quantum mechanics in non-relativistic case. Again stochastic diffusion processes are fundamentally time irreversible and dissipative. On the otherhand, the quantum processes are time reversible and believed to be non-dissipative in nature. Nelson proposed the hypothesis of background field so as to cause this sort of non-dissipative and reversible diffusion process responsible for the description of quantum particles. As he quoted in his list of problems, it still remains an open problem how to formulate the stochastic model of quantum mechanics with non-dissipative fluctuations. Several proposals [Ramanathan 1985]; [Bergia 1989] have already been made towards the explanation of this non-dissipative character of fluctuation.

Smolin [Smolin 1986] tried to analyze this situation based on the analysis of the Unruh effect [Unruh 1976]. His analysis also supports to the idea that gravitation or, more properly, metric fluctuations, are the right candidate for providing a physical background for stochastic mechanics. In a recent review, Bergia et al. [Bergia preprint] discussed extensively the fluctuation of space-time metric and nondissipative nature of quantum fluctuations.

A model of metric fluctuation in the spirit of Kaluza-Klein model can be constructed [Roy 1986]. In Kaluza-Klein model, there exists a dimension of length associated with this extra dimension. In Bergia et al. approach, if we take conformal metric fluctuation with ϕ as the random scalar field depending on the fifth co-ordinate x^5 in Kaluza-Klein scheme, then from the two point correlation

$$< \phi(x)\phi(y) >= L^2 f(x,y) \tag{8.1}$$

we can get the dimension of length L. This length can be obtained using the natural units like, $\hbar, c$ and G. Correspondingly, we can think of a mass as inversely related with this L. But this mass is almost of the order of Planck mass. In Nelson's mechanics the diffusion coefficient is related to the mass of the particle. So there remains ambiguity in the above Kaluza-Klein approach regarding fixing of the mass.

On the otherhand, Nelson's process can be generated by this type of conformal fluctuation of the metric. The fixing of the mass related to diffusion coefficient still remains an open problem. It may be mentioned that in the Chapter 3, (section 3.5), we have been able to show that the quantum potential can be generated out of the fluctuation of this fluctuating fifth dimension. Here we have assumed that if the elementary particle exists, space-time is distributed or fluctuated and the fluctuation is induced on the fifth dimension. Here, the length scale of fluctuation can be related to the mass of the elementary particle. The conditional expectation of the fluctuating component (fifth component) of the metric can be written as

$$\begin{aligned} Dg_{55} &= \lim_{\Delta t \to 0} \frac{1}{\Delta t} E(dg_{55}|q(t)) \\ &= \left(\frac{\hbar^2}{2m^2c^2}\right)\phi(x) \end{aligned} \tag{8.2}$$

$\hbar$, m and the function $\phi(x)$ arise on dimensional ground. Here, the average $< g_{55} >= \phi(x)$. Then, we can write

$$\sum_{\substack{M=1,2,3,4,5 \\ N=1,2,3,4,5}} g_{MN} dx^M dx^N = d\tau^2 + \eta_{55} dx^5 dx^5 \tag{8.3}$$

where $d\tau^2$ is the line element in 4-dimensional Minkowski space-time and $\eta_{55} =< g_{55} >= \phi(x)$.

Now the momentum in fifth dimension is

$$P_5 = \phi(x)\frac{dx^5}{d\lambda}; \quad P_\mu = U_\mu \frac{d\tau}{d\lambda} \tag{8.4}$$

with $\mu = 1,2,3,4.$ and U_μ are the covariant components of the ordinary velocity vector and λ is a parameter. For the light wave we can write

$$\frac{dP_5}{d\lambda} = 0; \quad \phi(x)\frac{dx^5}{d\lambda} = D \text{ constant.} \tag{8.5}$$

Again, $\frac{dx^5}{d\tau} = \frac{C}{\sqrt{\phi(x)}}$, (C is a constant). Then,

$$\phi(x)\frac{dx^5}{d\lambda} = \phi(x)\frac{C}{\sqrt{\phi(x)}}.\frac{d\tau}{d\lambda} \tag{8.6}$$

with

$$\frac{d\tau}{d\lambda} = \frac{E}{\sqrt{\phi(x)}}; \qquad D/C = \text{another constant } (E)$$

which gives

$$P_\mu = U_\mu\frac{d\tau}{d\lambda} = U_\mu\frac{E}{\sqrt{\phi(x)}} \tag{8.7}$$

Therefore the Hamiltonian can be written as

$$H = \frac{1}{2}\sum \eta_{MN}p^M p^N \quad M,N = 1,2,3,4,5. \tag{8.8}$$

From this Hamiltonian we obtain

$$\Delta\chi + \frac{4\pi^2}{c^2h^2}[h^2\nu^2 - m^2(c^4\phi(x)]\chi = 0 \tag{8.9}$$

where the total wave function can be written as,

$$\psi = \chi(x,y,z)e^{-2\pi i(\frac{x^5}{h} - \nu t)} \tag{8.10}$$

Then it coincides with relativistic Schrödinger equation with a potential term . No nonlinear term appears in this equation so that the system should be considered as conservative one. Here, though x^5 appears in the expression of ψ, it is immaterial for practical applications. Hence the quantum fluctuations generated due to this kind of metric fluctuation may be of non-dissipative in nature.

In section 3.5, we have mentioned that our approach can be generalized to non-zero spin particles [Ranganathan 1988]. So the above procedure may be generalized to construct Pauli-type equation within Kaluza-Klein framework and fluctuating space-time metric.

Some interesting proposals [Ramanathan, 1985] have been made to study the non-dissipative nature of quantum fluctuations using the idea of space-time metric fluctuation. Here, we shall start with Kolmogorov-Fokker-Planck equation for the spatio-temporal development of the transition probabilities governing the diffusion phenomena. The forward equation in a single space dimension can be written as

$$\begin{aligned} \partial_t p(x,t;x_o,t_0) &= \frac{1}{2}\partial_x^2[\alpha(x,t)p(x,t;x_0,t_0)] \\ &- \partial_t[\beta(x,t)p(x,t;x_0,t_0)] \end{aligned} \tag{8.11}$$

and the background equation as

$$\begin{aligned} -\partial_t p(x,t;x_0,t_0) &= \frac{1}{2}\alpha(x_0,t_0)\partial_{x_0}^2[p(x,t;x_0,t_0)] \\ &+ \beta(x_0,t_0)\partial_{x_0}[p(x,t;x_0,t_0)] \end{aligned} \tag{8.12}$$

where $p(x,t;x_0,t_0)$ is the transition probability of a particle in a diffusion process to be at space-time point (x,t) , α and β are the variance and the mean respectively. These are assumed to be relatively slow varying function of space-time point in comparison with the variations of the transition probabilities so that we may take

$$\alpha(x,t) = \alpha(x_0,t_0) = \sigma^2 \quad \text{(constant)}$$

$$\beta(x,t) = \beta(x_0,t_0) = \mu \quad \text{(constant)}$$

However, μ is dependent on external potential in describing quantum phenomena. But it will not violate the above condition. For example, an electron in the hydrogen atom treated as a Brownian particle must vary its position in space many orders of magnitude faster than its average drift in space, which is equivalent to μ . It is assumed that above diffusion is Markovian. So the transition probabilities p satisfy the Chapman-Kolmogorov conditions :

$$p(x,t;x_0,t_0) = \int_{-\infty}^{+\infty} p(x,t;x_0'',t_0'')p(x_0'',t_0'';x_0,t_0)dx_0'' \tag{8.13}$$

for $t \geq t_0'' \geq t_0$. We can define the probability density $p(x,t)$ in terms of the density $\rho(x_0,t_0)$ at (x_0,t_0) as

$$\begin{aligned} \rho(x,t) &= \int_{-\infty}^{+\infty} p(x,t;x_0,t_0)\rho(x_0,t_0)dx_0 \\ &= \{\phi(x,t)\}^2 \text{ for } t > t_0 \end{aligned} \tag{8.14}$$

and

$$\begin{aligned} \rho(x_0, t_0) &= \int_{-\infty}^{+\infty} p(x,t;x_0,t_0)\rho(x;t)dx \\ &= \{\phi(x_0,t_0)\}^2 \text{ for } t > t_0 \end{aligned} \tag{8.15}$$

where $\phi(x,t)$ and $\phi(x_0,t_0)$ are the real probability amplitudes at (x,t) and (x_0,t_0) respectively.

Now in terms of the amplitudes $\phi(x,t)$ and $\phi(x_0,t_0)$ we can write

$$\partial_t \rho(x,t) = \frac{1}{2}\sigma^2 \partial_x^2 \rho(x,t) - \mu \partial_x \rho(x,t) \tag{8.16}$$

or

$$\begin{aligned} \partial_t \phi(x,t) &= \frac{1}{2}\sigma^2 \{\partial_x^2 \phi(x,t) - \frac{2\mu}{\sigma^2}\partial_x \phi(x,t) + \Lambda^2 \phi(x,t)\} \\ &- \frac{1}{2}\sigma^2 \{\Lambda^2 - [\frac{\partial_x \phi(x,t)}{\phi(x,t)}]^2\}\phi(x,t) \end{aligned} \tag{8.17}$$

Taking $\Lambda = \frac{\mu}{\sigma^2}$, the above equation takes the form

$$\partial_t \phi(x,t) = \frac{1}{2}\sigma^2(\partial_x - \Lambda)^2 \phi(x,t) - \frac{1}{2}\sigma^2\{\Lambda^2 - [\frac{\partial_x \phi(x,t)}{\phi(x,t)}]\}\phi(x,t) \tag{8.18}$$

and similarly for backward equation. If we go over to a pure imaginary space-time (ix, it) by analytical continuation of the space-time (x,t) instead of the Euclidean space-time (x, it) of Euclidean quantum mechanics, we can rewrite the above equation for ϕ'

$$\begin{aligned} \partial_t \phi' &= -(i/2)\sigma^2 \exp(i\Lambda x)\partial_x^2(\phi' - \exp(-i\Lambda x)) \\ &- (i/2)\sigma^2 \left\{\Lambda^2 + [(\frac{\partial_x \phi'}{\phi'})]^2\right\}\phi' \end{aligned} \tag{8.19}$$

where $\phi' = \phi(ix, it)$.

Similarly the backward equation can also be constructed. Now it is evident that this equation is equivalent to Kolmogorov- Fokker- Planck equation in the analytically confined pure imaginary space-time in terms of the probabilty amplitudes, which no longer are real. Again

$$\text{Lim}_{(x_0,t_0)\to(x,t)}\phi'(x_0,t_0) = \phi'^*(x,t)$$

where ϕ'^* is the complex conjugate of ϕ'. But if we remain in real space-time, the Markov chain in the limit $(x_0, t_0) \to (x, t)$ would become singular in the sense that equations for $\phi(x,t)$ and $\phi(x_0, t_0)$ remain unrelated even at vanishingly small seperation of the Markov links. Therefore, we demand the invariance of the above equations for $\phi'(x,t)$ and $\phi'(x_0, t_0)$ under the phase transformations

$$\phi'(x,t) \leftrightarrow \phi'' \exp[i\Lambda x]$$

and

$$\phi'(x_0, t_0) \leftrightarrow \phi''(x_0, t_0) \exp[-i\Lambda x_0]$$

The invariance under these transformations imply that there will be no nonlinear "dissipative" terms under the condition :

$$|\Lambda|^2 \gg \left|[\frac{\partial_x \phi'(x,t)}{\phi'(x,t)}]^2\right| \cdot \left|[\frac{\partial_{x_0}, t_0)}{\phi'(x_0, t_0)}]\right| \tag{8.20}$$

Let us critically analyse this situation using Wigner's criteria for time reversal invariance. This demands that under the transformation

$$t \to -t, \; x \to x \text{ and } \phi'(x,t) \to \phi'(x,-t) = \phi'^*(x,t)$$

the equations for $\phi'(x,t)$ and $\phi'(x_0, t_0)$ should be invariant.

Applying these transformations to the equations for $\phi'(x,t)$ and $\phi'(x_0, t_0)$ we get nonlinear terms which do not allow the T-nonvariant phases as defined above. On the otherhand the application of the condition for $|\Lambda^2|$ allow us to have T-invariance. It is well known that T-invariance and nondissipativity are interrelated and hence the absence of the nonlinear term with the condition for $|\Lambda^2|$ allow us to construct a nondissipative diffusion equation. Thus using the phase transformations with the above constraint on $|\Lambda^2|$ we have

$$\partial_t \phi'' = -\frac{i}{2}\sigma^2 \partial_x^2 \phi'' - \frac{i}{2}\sigma^2 \Lambda^2 \phi'' \tag{8.21}$$

$$\partial_t \phi''(x_0, t_0) = \frac{i}{2}\sigma^2 \partial_{x_0}^2 \phi''(x_0, t_0) + \frac{i}{2}\Lambda^2 \phi''(x_0, t_0) \tag{8.22}$$

This equation has a formal similarity with the Schrödinger equation in ordinary space-time. For static diffusion process with $\partial_t \phi'' = 0$, we obtain

$$\partial_x^2 \phi'' + \Lambda^2 \phi'' = 0 \tag{8.23}$$

and the time independent Schrödinger equation is

$$\partial_x^2 \Psi + (\frac{2m}{\hbar})(E - V)\Psi = 0 \tag{8.24}$$

With

$$\Lambda^2 = (\frac{2m}{\hbar})(E - V)$$

So if the above equation for ϕ is to describe quantum phenomena a inverse length scale Λ should exist which must be the inverse de Broglie wavelength of the particle undergoing the diffusion. In case of time-dependent Schrödinger equation we have

$$\sigma^2 = \frac{\hbar}{m}$$

Thus we see that with this new interpreattion of the quantum phenomena as a particular linearized form of classical diffusion phenomena, the quantum nature is seen to due to the non vanishing of the variance σ^2 which in turn is dependent on the nonzero value of Planck's constant. The fluctuations of space-time metric may be a possible cause of quantum diffusion phenomena. This may help us to establish a rigorous framework to understand the relation between ordinary Brownian motion and the quantum diffusion phenomena. Here, the analytical continuation to pure imaginary space-time is only a convenient mathematical device to differentiate the stable (non-fluctuating) space-time metric from the fluctuating space-time metric at the microscopic scale.

In a recent paper [Percival, 1995] a complex diffusion in the comlex space-time plane has been considered to clarify the classical diffusion and quantum state diffusion. In such case, non-differentiable fluctuations in space-time on a Planck scale introduce stochastic terms into the equations for an existing alternative quantum theory; primary state diffusion. This theory, though is very attractive yet it is still in its infancy.

One of the attracting features of quantum fluctuations is their non-locality, in the sense that it is all pervading and it is the whole space-time that actually fluctuates. In modelling the space-time metric fluctuations as a candidate for quantum fluctuations, we have already discussed the significant role of conformal metric fluctuation both in quantum mechanics, cosmology and high energy astrophysics. It is important to study the momentum-energy properties of such conformal fluctuation and to estimate the velocity of propagation of disturbance for this kind of fluctuation. The principal idea behind this work is to show whether this kind of non-locality is related to the superluminal transmission of signal in the fluctuating space-time structure of vacuum.

For this purpose, let us take [Rosales 1991] the infinitesimal invariant interval for the fluctuating metric as

$$ds^2 = \exp[\alpha(x^\mu)]d\bar{t}^2 - \exp[\gamma(x^\mu)]d\sigma^2 \tag{8.25}$$

where $\alpha = \epsilon m\bar{t}$, $\gamma = \alpha + 2m\bar{t}$.

ϵ is the small parameter which specifies the fluctuation. The geodesic evolution of a particle is given by

$$\frac{d^2 x^\rho}{d\sigma^2} + \{\mu^\rho_\nu\}\frac{dx^\mu}{ds}\frac{dx^\nu}{ds} = 0 \tag{8.26}$$

For $\rho = 0$, it reduces to

$$\frac{d^2\bar{t}}{ds^2} + \frac{1}{2}(\dot{\alpha} + \dot{\gamma})(\frac{d\bar{t}}{ds})^2 - \frac{1}{2}\dot{\gamma}^2 = 0 \tag{8.27}$$

The dots denote the time derivatives. In the limit $\epsilon \to 0$, we get $(\bar{t} \to t)$

$$\frac{d^2 t}{ds^2} + m(\frac{dt}{ds})^2 - m = 0 \tag{8.28}$$

Then, $\frac{dt}{ds} = [1 + (Ae^{-mt})^2]^{1/2}$, A being a constant. For $\epsilon = 0$, the line element becomes

$$ds^2 = dt^2 - e^{2mt}d\sigma^2$$

and $ds^2 = 0$ indicates $c = c_0 e^{-mt}$, with c_0 as the velocity of light in Minkowski space-time. Then c may be considered as the velocity in the corresponding De Sitter manifold. Since $mt << 1$ even for cosmological scales, $c \approx c_0$ for all practical purposes.

Now let us define $u^2 = (\frac{dx}{dt})^2$.

Then

$$(\frac{ds}{dt})^2 = 1 - (u/c)^2 \text{ and } (Ae^{-mt})^2 = \frac{(u/c)^2}{1 - (u/c)^2} \tag{8.29}$$

Again let

$$\xi \equiv \frac{dt}{ds} = [1 - (u/c)^2]^{-1/2} \text{ and } \bar{\xi} = \frac{d\bar{t}}{ds} \tag{8.30}$$

The energy transmitted to the particle by the fluctuation is

$$\Delta\xi = Mc_0^2[\frac{d\bar{t}}{ds} - \frac{dt}{ds}]$$

where M is the rest mass. If we set

$$\frac{d\bar{t}}{ds} = \frac{dt}{ds} + \epsilon\phi$$

after simple calculations, we get

$$\epsilon\left[\frac{d\phi}{ds} + me^{mt}(\frac{dt}{ds})^2 + 2m\phi(\frac{dt}{ds}) - (\frac{m}{2})e^{mt}\right] = 0 \tag{8.31}$$

Again taking

$$\xi = [1 + (Ae^{-mt})^2]^{1/2},$$

after certain approximation we obtain the solution as

$$\phi = -\frac{1}{2}e^{mt}[1 + \frac{1}{2}(u/c)^2] + O[(u/c)^4] \tag{8.32}$$

The particle fluctuation energy can be calculated as

$$\delta E_f = -\frac{1}{2}\epsilon Mc_0^2 e^{mt}[1 + \frac{1}{2}(u/c)^2] \tag{8.33}$$

and the energy fluctuation of the vacuum gravitational field can be obtained as

$$\delta E_G = MV_g = \frac{1}{2}Mc_0^2\epsilon e^{mt} \tag{8.34}$$

where V_g can be obtained from the relation

$$\bar{g}_{00} = 1 + 2V_g/c_0^2 = 1 + \epsilon e^{mt} + O(\epsilon^2).$$

Thus for a static particle i.e., for

$$u = 0, \quad \delta(E_f + E_g) = 0 \tag{8.35}$$

This clearly indicates that the energy transformed by the vacuum to the particle is equal to the fluctuation in the potential energy of the gravitational vacuum field. This is valid at least for non-relativistic particles and at first order in the fluctuation parameter. This fact can be related to the important feature i.e., non-dissipating character of quantum fluctuation in the sense that they do not carry any dynamical quantities, and their non-locality can not thereby be used to transmit a "superluminal" signal.

Let us now calculate explicitly the propagation velocity of a signal in a fluctuating metric space. The propagation velocity for the particles with $u << c$ can be defined as

$$V(u) = \frac{\delta E(u) du}{d(\delta E(u))} = \frac{\delta E(u)}{d[\delta E(u)]/du} \tag{8.36}$$

This can be written explicitly using our previous formulae as

$$V(u) = \frac{c_0^2}{e^{2mt}u} + \frac{u^2}{2u} = \frac{c_0^2 e^{-2mt}}{u} + \frac{u}{2} \tag{8.37}$$

i.e.,

$$V(u) = \frac{c^2}{u} + \frac{u}{2} \tag{8.38}$$

This coincides with the non-relativistic limit of the phase velocity for a free particle, (whose group velocity is u), in de Broglie's theory :

$$V(u) = \frac{E}{p} = \frac{Mc^2 + (Mu^2)/2}{Mu} = \frac{c^2}{u} + \frac{u}{2} \tag{8.39}$$

It is clear from the above expression that this velocity is unphysical, in the sense that it does not describe any "physical" motion. This may be called as a kind of "information velocity". But one must be careful regarding the very notion of fluctuation velocity.

If we consider a particle even in a vacuum where gravitation field is negligibly small, the particle will fluctuate inspite of being in isolation. Then we can define a "fluctuation velocity" in a realistic sense. Now we can conceive of how the fluctuations propagate in momentum energy-space. It is also to be noted that this fluctuating velocity is equal to the phase velocity appearing in de Broglie's theory which might be help in finding a truly real-objective-source of the "primordial" quantum fluctuations, related to the conformal fluctuations of the space-time metric.

8.3. Limitations of the Measurement of Space-Time Distances :

One of the challanging problems in contemporary theoretical physics is posed by the incompatibility between quantum mechanics and (classical) general relativity. Probably the construction of a quantum theory incorporating gravity will require the development of a completely new understanding of geometry and physics. More insights can be obtained if the problems in which the incompatibility between them is more transparent. Works in this direction [Amelino-Camelia G., 1997] lead to the expectation that in quantum gravity unlike ordinary quantum mechanics, there might be bounds on the measurability of distances. The recent advances in string theory [Yoneya, 1989] also suggest that there exists a minimal length below which the space-time structure has no observable meaning. Infact, it is widely known that in any quantum theory of gravity, the Planck length puts a limitation on the smallest length scale of arbitrary measurement. But in most of these discussions it is not mentioned how this limitations on the measurement is reflected in observable quantities. Yonea [Yonea, 1989] critically analysed this situation in the context of conformally invariant string theory and proposed that the existence of a minimal length may be interpreted as a consequence of a new kind of dual indeterminacy relation. In all these discussions, the ultimate object of measurement is the space-time geometry. Wigner and his collaborators [Salecker , 1958] proposed that measuring space-time distances is possible using clocks only. Here, they avoided the use of measuring rods which are essentially macroscopical objects. For this purpose, we need a net of time-like geodesics as tight as possible. However, these geodesics are realized by real bodies (clocks) subject to quantum physics. For a length $s = cT$ a geodesic will develop a space-time uncertainty at least

$$\Delta x \approx \left(\frac{\hbar s}{Mc}\right)^{\frac{1}{2}} \tag{8.40}$$

where M is the mass of the body. Then geodesics become time-like world tubes rather than sharp world lines. Wigner imposed a plausible condition on the expected accuracy of measuring s :

$$\Delta s \approx \Delta x \tag{8.41}$$

i.e., space-like and time-like accuracies are identical. Now due to the finite size R of the clock travelling along the geodesic puts a limitation on the process of tightening the net. For $R > \Delta x$, the net is unnecessarily thin. So a suitable condition is

$$R \approx \Delta x$$

However, by increasing the mass M of the clock the accuracy of measuring the metric increases without limit. But, the masses distort the space-time just under measurement. This distortion leads to further deviation in measuring the length s, giving rise to new estimation

$$\Delta s' \approx (\frac{GM}{Rc^2})s$$

But a small distortion $\Delta s'$ needs large R. So in measuring the structure of the space-time we face two limitations :

- **(1)** Quantum uncertainty
- **(2)** Distortions as stated above

The first one decreases by taking a large mass M while the second one increases. We can get an optimum at

$$\Delta s \approx \Delta s'$$

An absolute limitation on defining the length of an individual geodesic has been considered as

$$(\Delta s)^2 \approx \Lambda^{\frac{4}{3}} s^{\frac{2}{3}}$$

Λ being the Planck length.

Taking this constructions, the optimal mass of the clock would be about $\hbar c^3 s/G^2$ which is emnormously large. This follows from the large size of the clock. Such a clock may measure quite well the length of an individual geodesic but it is hardly optimal for constructing a tight net structure [Diosi,1989].

Now let us consider the case when $R \approx \Delta x$. Again the localization of a free body along its world line can never be determined better than its Compton wavelength. Hence, in our optimal case

$$\Delta s \approx \frac{\hbar}{mc} \tag{8.42}$$

Now, with this condition, we have the absolute limit as

$$(\Delta s)^2 \approx (\frac{\Lambda^2 s}{R}) \tag{8.43}$$

via the optimal mass $(\frac{\hbar R}{GT})^{\frac{1}{2}}$. R may be considered as the cell size of the net. Formally the condition

$$\Delta x \approx (\frac{\hbar S}{Mc})^{\frac{1}{2}}$$

restricts the validity of the above equation (8.43) by limiting the length $(\Delta s)^2$ at measuring s is proportional to s. Hence, the succeessive independent measurements posses uncorrelated inaccuracies. Consequently, the above equation (8.43) remains valid for a world line of any length s since one can measure its length by successive measurements on shorter periods.This appears to be an inevitable uncertainty of measured lengths of time-like geodesics. Via Wigner's construction, this uncertainty propagates into the space-time geometry which therefore can not be sharply measured. So there is no possibility to consider it as sharp. It is now possible to calculate the uncertainty of the metric tensor [Diosi, 1989]. Diosi gave an estimate for fluctuations $g_{\mu\nu}$ whose rigorous determination may play important role in the formulation of relativistic quantum gravity.

It is interesting to note that in the framework of critical string theory we can write a modified uncertainty relation

$$\delta x \delta P = \hbar + \frac{L_p^2}{\hbar} \delta P^2 \tag{8.44}$$

Amelino-Camelia [Amelino-Camelia, 1997] suggested a measurability bound by the observation that once the gravitational effects are taken into account, it is no longer possible to rely on the availability of classical agents for the measurement procedure. Based on this observation we may arrive at the bound

$$\min[\delta L] = \sqrt{\frac{L L_p^2}{s}} \tag{8.45}$$

where s is the length scale associated with spatial extension of the clocks. This bound is always large than L_p for $L_p <$ or $\approx s <$ or $\approx L$ and is maximal with idealized scenario $\sim L_p$, in which

$$\min[\delta L] = \sqrt{L L_p}$$

Then we can write

$$\delta x \ \delta P = \hbar + \frac{c L_p^2}{\hbar} . \frac{T^*}{S} \delta P^2 \tag{8.46}$$

where T^* is a time scale characterizing the process of observation of the system.

But certain structures encountered in quantum K-Poincaré group lead to enlarged measurability bounds of the type $\{\min[\delta L] = \sqrt{\frac{LL_p^2}{S}}\}$ although the deformed uncertainty relation responsible for the bound is not of the type(8.46).

In quantum $K-$ deformed Minkowski space

$$[x_j, x_k] = 0, \qquad [x_j, t] = \frac{x_j}{k}$$

can be interpreted as implying that the unceratinties on x_j and t satisfy

$$\delta x_j \delta t \geq \frac{x_j}{|k|}$$

Then we can find the implications for measuring bounds of this deformed uncertainty relation by analyzing simple procedures for the measurement of the distance L between two bodies.

After simple calculations it is found

$$\min[\delta L] \sim \sqrt{\frac{cL}{|k|}} \tag{8.47}$$

This produces the realtions

$$\min[\delta L] = \sqrt{LL_p} \tag{8.48}$$

by appropriate association of the scale k to the Poncaré scale. But it is necessary to clarify the physical meaning of the algebraic concept of non-commutative coordinates as proposed above.

8.4 Relativistic Cosmology and Time Singularity :

In recent time new interests have arisen to study the region near Planck length. We have already discussed the interesting developments of covariant string field theory and the role of minimal length. The length scale of space-time fluctuations has been shown to be of the order of 10^{-32} cm [Blokhintsev, 1975] which is slightly larger than Planck length. In most of these works it has been emphasized that even the metrizibility of the space-time might be lost below Planck length. Belinskii, Lifchitz and Khalatnikov studied the singularity of the metric which has oscillatory character using the model of universe with a homogeneous space of type IX [Belinskii,1982]. They pointed out the existence of an oscillatory regime near time singularity of the metric within the framework of general theory of relativity (GTR). We know that that this time singularity in cosmological solution of the Einstein equations raises some fundemental issues like : to what extent the existence of a time singularity is a necessary general property of cosmological models, or whether it is really caused by the specific simlpifying assumptions on which the models are based. If the presence of the singularity were independent of these assumptions, it would mean that it is inherent not only to special solutions, but also to the general solution of the Einstein equations. Thus their result has great significance not only in GTR but also in studying the origin of the fluctuations of the space-time metric itself from within a very general framework. Starting with a model of universe in a homogeneous space of type IX, let us concentrate our discussions on the behavior of this model near the singularity (which we choose at the time of origin $t = 0$). However, the presence of matter does not affect the qualitative properties of this behavior. Let us assume first that the space is empty.

Now in homogeneous space, the line element can be written as

$$dl^2 = \gamma_{\alpha\beta}(x^1, x^2, x^3)dx^\alpha dx^\beta \tag{8.49}$$

Then after transformation, the space line element becomes

$$dl^2 = \gamma_{\alpha\beta}(x^{1'}, x^{2'}, x^{3'})dx^{\alpha'} dx^{\beta'} \tag{8.50}$$

with the same functional dependence of $\gamma_{\alpha\beta}$ on the new coordinates.

Here, as space is considered as three dimensional, the different transformations of the group are labelled by three independent parameters. In Euclidean space, the homogeneity of space is expressed by the invariance of the metric under parallel displacements of the Cartesian coordinate system. Each translation is determined by three parameters : the coordinate of the displacement vector of the coordinate origin. In case of non-Euclidean homogeneous space, the transformations of its group of motions leave invariant three independent linear differential forms, which do not, however, reduce to total differentials by any coordinate functions. Let us write these forms as

$$e_\alpha^{(a)} dx^\alpha$$

in which the labels (a) indicate three independent vectors (coordinate functions). Let us denote these vectors a -frame.

Now, we can construct a spatial metric invariant under the given group of motions :

$$dl^2 = \eta_{\alpha\beta}(e_\alpha^{(a)} dx^\alpha)(e_\beta^{(b)} dx^\beta) \tag{8.51}$$

with the metric tensor $\gamma_{\alpha\beta} = \eta_{ab} e_\alpha^{(a)} e_\beta^{(b)}$.
Here, the coefficients η_{ab} symmetric in the indices a and b, are functions of the time.

Considering the diagonal $\eta_{ab}(t)$, the spatial metric is written as

$$\gamma_{\alpha\beta} = a^2 l_\alpha l_\beta + b^2 m_\alpha m_\beta + c^2 n_\alpha n_\beta \tag{8.52}$$

where a^2, b^2, c^2 are three diagonal elements and $\vec{l}, \vec{m}, \vec{n}$ are frame vectors. For a space of type IX, the structure constants are

$$C^{11} = C^{22} = C^{33} = 1; \quad C_{23}^1 = C_{31}^2 = C_{12}^3 = 1$$

Now considering these constants and a diagonal matrix η_{ab}, components of Ricci tensor $R_{(a)}^0$ vanish identically in the synchronous reference system. In fact the components of the Ricci tensor can be written as

$$R_0^0 = -\frac{1}{2}\dot{\chi}_{(a)}^{(b)} - \frac{1}{4}\chi_{(a)}^{(b)}\chi_{(b)}^{(a)} \tag{8.53}$$

$$R_{(a)}^0 = -\frac{1}{2}\chi_{(b)}^{(c)}\left(C_{ca}^b - \delta_a^b C_{dc}^d\right) \tag{8.54}$$

$$R_{(a)}^{(b)} = -\frac{1}{2}\sqrt{\eta}\left(\sqrt{\eta}\chi_{(a)}^{(b)}\right)' - P_{(a)}^{(b)} \tag{8.55}$$

where the dot(.) denotes differentiation with respect to t. It clearly shows that even the nondiagonal components $P_{(b)}^{(a)}$ vanish when

$$\begin{aligned} P_{(a)}^{(b)} &= \frac{1}{2\eta}\{2C^{bd}C_{ad} + C^{db}C_{ad} + C^{bd}C_{da} \\ &- C_d^d(C_a^b + C_a^b) + \delta_a^b[(C_d^d)^2 - 2C^{df}C_{df}]\} \end{aligned} \tag{8.56}$$

The Einstein equations give the following system of equations for the functions a, b, c :

$$\frac{(\dot{a}bc)^{\cdot}}{abc} = \frac{1}{2a^2b^2c^2}[(b^2 - c^2)^2 - a^4] \tag{8.57}$$

$$\frac{(a\dot{b}c)^{\cdot}}{abc} = \frac{1}{2a^2b^2c^2}[(a^2 - c^2)^2 - b^4] \tag{8.58}$$

$$\frac{(ab\dot{c})^{\cdot}}{abc} = \frac{1}{2a^2b^2c^2}[(a^2 - b^2)^2 - c^4] \tag{8.59}$$

$$\frac{\ddot{a}}{a} + \frac{\ddot{b}}{b} + \frac{\ddot{c}}{c} = 0 \tag{8.60}$$

This system of equations takes a simple form if we introduce the logarithms of a, b, c as α, β, γ in place of the functions a, b, c i.e.,

$$a = e^{\alpha}; \quad b = e^{\beta}; \quad c = e^{\gamma}; \quad dt = abcd\tau \tag{8.61}$$

Then

$$\begin{aligned} 2\alpha.\tau.\tau &= (b^2 - c^2)^2 - a^4 \\ 2\beta.\tau.\tau &= (a^2 - c^2)^2 - b^4 \\ 2\gamma.\tau.\tau &= (a^2 - b^2)^2 - c^4 \end{aligned} \tag{8.62}$$

and

$$\frac{1}{2}(\alpha + \beta + \gamma)_{.\tau.\tau} = \alpha.\tau^{\beta}.\tau + \alpha.\tau\gamma.\tau + \beta.\tau^{\gamma}.\tau \tag{8.63}$$

where the subscript $.\tau$ denotes differentiation with respect to τ. Again

$$\alpha.\tau.\tau + \alpha.\tau^{\gamma}.\tau + \beta.\tau^{\gamma}.\tau = \frac{1}{4}(a^4 + b^4 + c^4 - 2a^2b^2 - 2a^2c^2 - 2b^2c^2) \tag{8.64}$$

The above set of equations (8.57 - 8.60) can not be solved analytically. But we can study it qualitatively in the neighorhood of the singular point. It is found that [Belinskii, 1982] the process of evolution of the model towards the singular point is made up of successive series of oscillations, during which the distances along two of the space axis oscillate, while they fall off monotonically along the third. The volume dropss off according to a law that is close to $\sim t$. In going from the one series to the next, the direction along which there is a monotonic drop off of distance, shifts from one direction to another. Asymptotically the order of these shifts takes on the charachter of a random process. The order of succession of lengths of successive series of oscillations also takes on stochastic character. Let the minimal value of the parameter u is $u = k_0 + x_0$, k_0 being an integer and $x_0 < 1$. The length of the first series of oscillations will be k_0, while the initial value for the next series will be $u_1 = \frac{1}{x_0} = k_1 + x_1$ etc. From this it is easy to conclude that the lengths of successive series will be given by the elements $k_0, k_1, k_2, \cdots$ of the expansion of u_0 in infinite (for irrational u_0) continued fraction

$$u_0 + k_0 + \cfrac{1}{k_1 + \cfrac{1}{k_2 + \cfrac{1}{k_3 + \cdots}}}$$

The succession of values of further elements in such an expansion is distributed according to statistical laws.

As we approach the singularity, the successive series of oscillations crowd together. An infinite number of oscillations are contained between any finite world time t and the moment $t = 0$. The natural variable for describing the time behavior of this evolution is not the time t but its logarithm lnt, in terms of which the whole process of approach to the singular point is spread out to $-\infty$.

It raises interesting possibility to look into the connection of the fluctuation of metric and this oscillatory regime near time singularity.

It is worth mentioning that in a series of recent papers, Percival [Percival 1995] studied a series of the non-differentiable fluctuations in space-time on a Planck scale. In this frame work, non-differentiable fluctuations in space-time give rise to stochastic terms into the equations for quantum states. This proposal leads to new foundation for an existing alternative quantum theory : Primary State Diffusion (PSD). PSD theory is developed from quantum state diffusion theory in which both quantum and classical behavior are derived from the same dynamics. Unlike the earlier quantum state diffusion theory, here, the diffusion in PSD is independent of any division between system and environment or observed system and observer. Before going into the PSD, let us discuss the structure of quantum state diffusion [Gisin, 1993].

The basic idea behind this approach is that evolution of a quantum system is determined through a unique correspondence between Bloch equation and an Ito diffusion equation for normalized state vector $|\psi>$ of an individual system of the ensemble. If H is the Hamiltonian for an open system and L_m are the environment operators which represents the interaction of the system with its environment, then the Bloch equation is

$$\dot{\rho} = -\frac{i}{\hbar}[H,\rho] + \sum_m (L_m \rho L_m^+ - \frac{1}{2}L_m^+ L_m \rho - \frac{1}{2}\rho L_m^+ L_m) \tag{8.65}$$

The corresponding quantum state diffusion equation is a stochastic differential equation for the normalized state vector $|\psi>$, whose differential Ito form is given by

$$\begin{aligned} |d\psi> &= -\frac{i}{\hbar}H|\psi> dt + \sum_m (< L_m^+ >_\psi L_m - \frac{1}{2}L_m^+ L_m \\ &- \frac{1}{2} < L_m^+ >_\psi < L_m >_\psi) |\psi> dt \\ &+ \sum_m (L_m - < L_m >) |\psi> d\xi_m \end{aligned} \tag{8.66}$$

where $< L_m >_\psi = <\psi | L_m | \psi>$ is the expectation of L_m for state $|\psi>$ and the density operator is given by the mean over the projectors on to the quantum states of the ensemble,

$$\rho = M|\psi><\psi|$$

M being mean over the ensemble.

In the right hand side of the above equation (8.60) for $|\psi>$ the second term represents the drift of the state vector while the 3rd term is due to random fluctuations. Both the terms may be due to the interaction of the system with its environment. The random variables $d\xi_m$ are assumed to satisfy the following relations

$$\begin{aligned} M(d\xi_m) &= 0 \\ M(d\xi_n\, d\xi_m) &= 0 \\ M(d\xi_n^* d\xi_m) &= \delta_{nm} dt \end{aligned} \tag{8.67}$$

This is a complex Wiener process which is normilized to unity while the independent real and imaginary Wiener processes are each normalized to 1/2. Like Bloch equation state diffusion (SD) is valid for sufficiently short time for the interaction of the system with its environment. However, they are both dependent on the choice of the boundary between the system and its environment. In many cases, this choice can be made such that the dependence is weak. It appears from the above ideas that the SD theory of quantum mechanics is analogous to the classical circuit theory. An electrical circuit includes both Hamiltonian elements, with capacitance and inductance, and non-Hamiltonian elements, such as resistors, which represents the interaction of the system with its environment.

In quantum state diffusion theory, the following principles are usually considered to be valid :

1. Interactions with independent parts of the environment lead to statistically independent quantum state diffusions.
2. There is an interaction with the environment in which the oveall probability of being in a subspace of the state space is conserved, the corresponding amplitudes of the state of individual system of the ensemble fluctuates, unless the effective rate of interaction of the system with its environment is zero.
3. Probability conservation and fluctuation together imply localization.
4. Localization of states in position space follows from locality of interactions. The rest frame of the localization is determined by the environment.

Localization theorems have been proved by Gisin et al. [Gisin, 1993]. The theorems are known as dispersion entropy theorem and the general localization theorem. Later on Percival [Percival, 1994] generalized these ideas and now known as Primary State Diffusion (PSD). The distinguishing feature of PSD is that the diffusion of quantum states, which dominates on very small time scales, is primary and the usual Schrödinger evolution of the state vector is derived from it. A time constant τ_0 is introduced in PSD which is supposed to be approximately the Planck time

$$\tau_p = (\hbar G c^{-5})^{\frac{1}{2}} \simeq 5 \times 10^{-44}\text{sec}$$

This suggests localization of wave packets of Caesium atoms over a distance of 15au, probably within the range of feasible matter interferometry experiments. Initially PSD theory was derived without any reference to the structure of space-time. More recently this theory has been related to the fluctuations of space-time [Percival, 1995].

The space-time PSD is based on a stoachastic differential structure and a fluctuating space-time metric. These are taken as quantum fluctuations but they are expressed in terms of non-differentiable relations between classical space-time. Basically it is made of two parts : The properties of a Minkowski universe and its Lorentz frames and the properties of our universe that are obtained by local non-differentiable space-time transformation between the Minkowski universe and ours. As in general Relativity (GR) our universe can be locally related to a Minkowski universe by a space-time transformation, even down to the Planck scale and below. But whereas, on a macro scale the space-time of our universe and the transformations appear to be smooth and differentiable, so on the Planck scale the space-time and the transformations are fluctuating and non-differentiable.

These issues should be studied in great depth so as to have deep insight to understand the role of quantum fluctuations as well as the structure of space-time at small distances.

EPILOGUE : NEW DIALOGUE OF MAN WITH NATURE

"The world thus appears a complicated tissue of events, in which connections of different kinds alternate and overlap or combine and thereby determine the texture of the whole"

Werner Heisenberg : Physics and Philosophy

9. EPILOGUE : NEW DIALOGUE OF MAN WITH NATURE

We have presented an overall and integrated view of fluctuating or statistical geometry and the foundational problems at different levels of understanding in Microphysics and Cosmology. The reader should be aware that our view is by no means conclusive, however, and that the study of the fuzziness in geometry is still in its infancy. We suggest a need for revolution in our conceptualization of space-time as envisaged by Edward Witten [Witten 1996] with great excitement:

"Contemporary developments in theoretical physics suggest that another revolution may be in progress, through which a new source of "fuzziness" may enter physics and space-time may be reinterpreted as an approximate, derived concept".

9.1 : FUZZINESS IN SPACE-TIME

The adjective "fuzzy" becomes very popular [Zadeh 1965] in the contemporary investigations, specially, concerning the logical and set theoretical foundations of mathematics. One of the main reasons for the rapid development of this notion emerges from the realization that the world around us is full of uncertainty, the information we obtain from the environment and the data resulting from observations or measurements are, in general, vague and incorrect. In fact, the 19th century, which marks the beginning of the modern age of science, is an era of the conceptual developments in the art of measurement. Till the beginning of this century, it was believed that through careful design and ample data, the error in any measurement could be made arbitrary small. The birth of quantum mechanics shattered this belief. Here, the uncertainties of the measurements are inherent in the measurement process itself and in principle it cannot be removed. Again the rapid developments in modern computing machine and the theory of information processing in the field of speech analysis, image processing and pattern recognition, complex decision problems and in general, what is included in the field of artificial intelligence [Dattamajumdar 1993], expedite the process to formalize the notions like fuzzy sets, fuzzy ordering etc. Briefly speaking, every formal description of the real world or some of its aspects is, in every case, only an approximation and idealization of the actual state. Albert Einstein made a remarkable statement [Einstein 1921] regarding the physical world and reality :

" As far as the laws of mathematics refer to reality, they are not certain; and so far as they are certain, they do not refer to reality".

The very notion of fuzzy sets enable us to handle and to study the degree of such kind of uncertainty in a purely mathematical and formal way.

DEFINITION :

Let X be a non-empty set. A fuzzy set A in X is a pair $< X, f_A >$ where f_A is a function defined on X and taking its values in the set $< 0, 1 >$ of reals. f_A denotes the degree or the likelihood of the membership of this element in the considered fuzzy set. If f_A takes only 0 or 1, it reduces to the characteristic function in the "classical" sense. Therefore, f_A is usually called the (generalized) characteristic function of the fuzzy set A in X. On the other hand we have also tried to expose a kinds of "fuzziness" in the space-time structure throughout this monograph. It seems that we need two kind of fuzziness in explaining the natural phenomena though similar connotation is used in both the cases in its literal sense. Karl Menger [Menger 1951] who began his systematic exposition of statistical geometry long before the above mentioned type of fuzzy set theory, intended to describe also the real world or some of its aspects in the spirit of Einstein. On the very intuitive level, the notion of the statistical and the notion of fuzzy metric might have a deeper connection.

In chapter 1, we have already mentioned and discussed that Kramosil and Michalik [Kramosil 1975] studied the notion of fuzziness in the context of metric spaces and they introduced the notion of "fuzzy metric". This notion of "fuzzy metric" is being compared with that of statistical metric space and both the conceptions are proved to be equivalent in a certain sense. These authors emphasized that there is an intuitive difference between probability and fuzziness, even if both of these notions wished to describe some aspects of uncertainty connected with events and notions in the surrounding world. Introducing the fuzziness in topological spaces may help us to understand this difference in a more explicit way. Then a unified description may be possible to explain the physical phenomena in various strata of knowledge.

Recently, a new mathematical structure [Busch 1991] for observables in quantum mechanics has been developed using the concept of fuzzy set theory within the axiomatic framework of Hilbert space. At first let us designate the membership function in fuzzy set theory as f_A and the characteristic function in the classical sense χ_A. The main difference of f_A from the characteristic function χ_A is that f_A may assume any value in the real interval $[0,1]$. Now we proceed with the corresponding change in the description of the observables. In Hilbert space formulation of quantum mechanics, the description of a physical system S is based on a Hilbert space H (complex, separable) with inner product $< .|. >$. Let $L(H)$ denotes the set of bounded operators on H. A spectral projection is defined as

$$Q(A)\phi(q) = \chi_A(q)\phi(q)$$

This is known as projection value (PV) measure in quantum mechanics. It can be identified with a unique self-adjoint operator acting in H. The observable in quantum mechanics is usually represented by self adjoint operator. Now if we replace χ_A by f_A, we get a new "effect"

$$(a(A)\phi)(q) = f_A(q)\phi(q)$$

where, $a : A \rightarrow a(A)$ is a positive operator value measure(POV) called Unsharp Position Observable or fuzzy observable.

It is now clear thatPV measure can be treated as a special case of POV measure. We call the observable corresponding toPV measure as sharp observable and unsharp or fuzzy observable corresponding to POV measure. Here, one can construct the unsharp spin observables which have great impact on Bell's theorem, non-locality and complementary principle. This framework is not just a mathematical extension of Hilbert space formalism of quantum mechanics. It is very much useful to model the real physical situation and its effect in a mathematically consistent way. In this framework, unsharpness or fuzziness is introduced at two levels. Unsharpness is introduced in preparing the state of the object and is also involved at the level of measurement. These double roles of the state vector have a profound impact on the consistent interpretation of Heisenberg uncertainty relation.

A similar situation arises also in the context of signal processing, image analysis and pattern recognition [Roy 1996]. In the domain of pattern recognition and image processing, some sort of ambiguity arises due to the background. A gray tone image possesses ambiguity within pixel because of possible multi valued levels of brightness in the image. This indeterminacy is due to inherent vagueness rather than randomness. Incertitude in image pattern can be explained in terms of grayness, spatial ambiguity or both. Grayness ambiguity means indefiniteness in deciding whether a pixel is white or black. Spatial ambiguity refers to indefiniteness in shape and geometry of the region within the gray level of the image. The present author along with his collaborators [Roy 1996] have shown that in the field of image processing and computer vision at one level, the background introduces uncertainty like the unsharpness introduced in preparing the state of the object in quantum mechanics and uncertainty is also involved at the level of measurement. This has a striking similarity with the two roles of the state vector as already mentioned in the context of unsharp observables in quantum mechanics.

It is interesting to note that we are getting fuzzy structure in different strata of knowledge and this fuzziness seems to be inherent in nature. In this monograph, we have concentrated our discussion on C-spaces which seem to play significant role in microscopic as well as in cosmological domain. Here, the clouds do not have fixed boundaries and hence known as hazy lumps. In sixties or early seventies people thought them to be seats of elementary particles. Now due to the development of string theory, these hazy lumps may be considered due to string structure. So, the space-time should be considered as derived concept.

In the above mentioned C-spaces, the Frechet-Minkowski metrics $d_{(\beta)}$ [Schweizer 1983] have a remarkable structure. The Frechet-Minkowski metrics $d_{(\beta)}$ is defined as

$$d_{(\beta)}(p,q) = \left\{ \begin{array}{cc} m_{(\beta)}F_{pq} & \beta \in (0,1] \\ \left(m^{(\beta)}F_{pq}\right)^{1/\beta} & \beta \in [1,\infty) \end{array} \right\}$$

Here, $m^{(\beta)}$ can be written as

$$m^{(\beta)}F = \int_{R+} x^{\beta} dF(x)$$

At small scale, they are non-Euclidian. In fact, the distance between any two distinct points p, q is not smaller than a fixed positive multiple of σ_{pq}. In the large, they are asymptotic to the Euclidean metric of R^n itself. Intuitively we can think of the following situation. If we associate space of particles or string like objects to the clouds, that may move about, we see on the one hand, that, as the Euclidean distance $d(C_p, C_q)$ between the centers of the clouds p and q approaches zero, any Frechet-Minkowski distance between p and q remains greater than a positive number; and on the other hand that, when it is large, $d(C_p, C_q)$ is a good estimate of the distance between two objects. This result seems to play significant role in constructing a unified description of matter and space-time from the scale of string size or length of fluctuation of metric (of the order of 10^{-32} cm) to the cosmological length scale.

9.2 : SPONTANEITY, ORDER AND WHOLENESS

We have already mentioned that Einstein [Einstein 1921] discussed the possibilities of two kind of geometries:**axiomatic geometry** and **physical geometry**. The discovery of General Theory of Relativity led Einstein to introduce the idea of physical geometry. However, Bunge [Bunge 1977] discussed the possibilities of three kind of geometry:

- **Mathematical**
- **Physical**
- **Ontological**

Briefly speaking, mathematical geometry is the collection of theories defining spaces of all kinds, where the space is defined as an arbitrary set with a minimal structure, namely a topology. This structure may or may not depend upon the nature of elements of set.

A physical geometry is a theory where things and events are represented in physical space (or in space time). In this case, the mathematical geometry is enriched with semantic assumptions i.e. "correspondence rules" specifying the referents of the theory (i.e., light rays) and what properties of such reference (i.e., their mutual separation) those concepts represent. For example, a triangle could be determined by three intersecting light rays. Here, the objects of a physical geometry are physical objects and as such subject to measurement at least in principle.

In the third category i.e., in ontological framework, space time is considered to be collection of facts with their separations, where, "separation" is constructed in terms of facts only. Bunge called it as **Chronotopics**. In fact, chronotopics makes use of both ontology and pure mathematics and it intends to underpin physical geometry. These different views on the nature of space-time can be described in more formalized language as: the container view, the prime stuff and the relational views.

(a) The Container View :

Here, the space and time are described in purely mathematical terms like a supreme container which cannot be described in physical terms. According to this view, the space time (i.e., container) exists by itself even if the physical objects or things do not exist. Einstein shattered this view in formulating the general theory of relativity where the very structure of space-time is influenced by things contained (though not determined). Moreover, this doctrine is philosophically unsatisfactory as it does not really answer - except metaphorically - the question "what are space and time?". Bunge rightly pointed out "it is particularly unacceptable to any ontology which, like ours, has no room for objects that are neither things nor properties of things nor relations among things".

(b) The Prime Stuff View :

According to this view, the space and time are just as absolute as in the container view, insofar as their properties depend on nothing. However, the difference of the container view and prime stuff view lies in the fact that here, things are not in space time but instead the later constitutes the former. For example, in Wheeler's vision [Wheeler 1962], an electrically charged particle is said to be identical to a "worm hole" in the basic space time manifold. Indeed, every physical object is regarded as a local warping of the space time. Though this view is very fascinating, it is unfortunately semantically and ontologically too simple to be true.

(c) The Relational View :

The relational view has been proposed and discussed in various contexts by thinkers as diverse as Aristotle, Lucretius, Augustine, Leibnitz, Lamberte, Lobachevski, Riemann, Engles, Mach, Whitehead, Rob, Van Dantzig, Fokker, Menger and Penrose. In Leibnitz's words: "Space is an order of possible coexistence and time an order of successive". According to this view, space and time do not exist by their own. This should be considered as a network of relations among factual items/things and their changes. The mathematical properties like topological, affine or metrizability is not inherent in space- time itself but due to relations among things or events. Ontologically one should say that space and time are not self existing (absolute objects of uncertain ontological status; neither things nor their properties). Space time should be considered as the basic structure of the totality of all possible facts.

This ontological geometry seems to be compatible with the recent developments of elementary particle physics at very high energy. In chapter 1, we have mentioned that the recent high energy experiments reveal the existence of a minimal length of the order of 10^{-17} cm below which no measurement is possible. So, the spatial continuum may have to be replaced by a granular or quantized space with a fundamental length built in it. But this hypothesis does not solve Poincaré dilemma of distinguishing between a transitive mathematical and intransitive physical relation of equality,

"It does seems much more helpful to retain the transitive relation in mathematics and to introduce for the distinction of physical and physiological quantities as probability, that is, a number lying between 0 and 1".

Now if we consider the geometry where points are not the primary entities but rather the lumps and introduce the concept of probability so that same two objects are sometime treated as identical and sometimes as distinguishable, then it is possible to construct an alternate geometry which is more compatible with the physical reality. Here, lumps do not have sharp boundaries. These are considered as hazy or fuzzy lumps which are undefined concepts, whereas points appear as the result of limiting or intersectional processes applied to these lumps.

In this new geometry, we have introduced two basic concepts :

- **Concept of hazy or fuzzy lumps**
- **Statistical notion.**

These hazy lumps can be identified with the extended structure of elementary particles or may be comparable with the size of string-like objects. Now the statistical elements enter either due to the existence of random zero point field or stochastic nature of string-like objects. At this point, we face deep philosophical issues like the assignment of mathematical properties as topological or metrical to the above kind of space (i.e., space as a set of hazy lumps). The adjective **topological** recognizes the need to introduce a notion of order whereas the notion of metric is related to the notion of measure. The notion of order is more or less known to everyone. This has been used in various contexts in a more wide sense.In the context of ontological geometry, this order may be related to some regular arrangement of objects or forms in lines or in rows. Rather we shall consider much more general concept of order ; for example : the order of growth of living being, the order of living species, the order of society, the order of a musical composition, the order of paintings, the order which constitutes the meaning of a communications etc.

One may ask the general question "**What is order ?**" However , there exists a difficulty in defining the notion "**order**", as the very meaning of "definition" is in itself an ordered way of thinking and using language. David Bohm [Bohm 1971] made a detailed analysis regarding this question. He used a term called the degree of order. In his framework , the term "**disorder**" is no longer used,instead we call it as an order of indefinitely high order. For example, let us consider the random curves - such as the trajectory of Brownian particle This trajectory is a continuous but no where differentiable function. So one can associate an indefinitely high degree of order with such a "random" curve. But it should be noted that order is not to be identified with predictability. Predictability is a special kind of order such that a few steps determine the whole order (as in curves of low degree). There may be complex and subtle orders which are not in essence related to predictability. An image or painting may be highly ordered but this order does not permit one to be predicted from another.

Now as soon as we introduce the notion of degree of order, we have tacitly imported the idea that each suborder has a limit. Again the primitive meaning of the word "measure" is to limit or "boundary". As a result, by putting together the notions of "order" and "measure" a new notion called "structure" may emerge in more and more complex contexts. In case of a geometrical figure like triangle,it is a structure formed by the lines where each of these lines is in itself an order of segments, which is limited (i.e. measured) by other lines. So the notion of structure implies a harmoniously organized totality of orders and measures, which is both hierarchic (i.e. on more levels) and extensive (spreading on each level).The principle of structure is universal. Living beings are in continual movement of growth and evolution of structure which is highly organized. In physics, matter is considered to be consist of atoms, liquids or gaseous structures.They make larger structures, going on up to planets, stars, galaxies etc. If we study the history of physics and the emergence of new order, it is found that it has been a tendency to regard certain basic notion of order as permanent and unchangeable. Then the task of the physics is to accommodate new observations by means of adaptations within basic notions of order, so as to fit the new facts. This accommodation is considered to be the main activity in physics and the emergence of new order is a rare phenomena when the normal process of accommodation breaks down. Then what is needed is a fresh perception of the "whole fact". This includes not only the results of experiments, but also the failure of certain lines of theory to fit the experimental results in a **common measure**. Then, one has to be very sensitively aware of all the relevant differences which underlie the main orders in the old theory to see whether there is room for a change of overall order.

In itself, the above concept of "**wholeness**" is not new. There is in the **Rig Veda**, the oldest human writing in existence, a beautiful passage describing creation from such kind of "wholeness". This concept of creation has been elaborated in Indian Philosophy in various contexts (in Vedas and Upanishads). There are two ingredients in this philosophy :

- **The idea of spontaneity**
- **The concept of extension (instead of points or point like-things).**

We read in "**Mundaka Upanishad**" :

"As the Urnanabhi (spider) spins the thread out of its own body....even so that the whole Universe has come out of the Imperishable being".

Here lies the concept of spontaneity in creating the Universe from one or undivided whole. The same reality which is called in its individuality (microcosmic) aspect as **Atman** is called **Brahman** in its collective (macrocosmic) aspect. Even etymologically, Brahman means Big, Vast, Immense. There does not exist any dualism between singularity and multiplicity in the concept of "wholeness" as multiplicity or plurality is unknown. There is only one thing and that, what seems to be plurality, is merely a series of different aspects of this one thing, produced by deception (called **Maya** or Illusion in Vedanta). In Hindu Cosmology (Vivekananda 1900), it is said that **Brahman** is the only supreme reality when **Prana** (primal energy) and **Akasha** (aether-like thing) are merged. **Prana** is almost inseparable from **Akasha** but there is no vibration or motion at this stage. It seems, infinite degrees of freedom are completely frozen. Then by **Being Will** , there is vibration, condensation happens and thence matter. But this **Being Will** is also in the **Undivided wholeness** as the concept of plurality does not exist in this **wholeness.** Now this **Primal** vibration can be easily identified with the spontaneous fluctuation.

Subsequent to the above passage, we read:

"Brahman became self extended by knowledge of creation....a stream of derivatives now follow: Hiranyagarbha, mind, five elements, from those substances rise, the world order and thence the fruits of action follow".

Here, **Brahman** becomes extended by itself spontaneously. So space may be thought of as self extension of **Brahman.** This extension has been compared with a string-like thing. Infact, **Hiranyagarbha** is called Sutra-atman, Sutra meaning string. The Bhagavad-gita declares that the Supreme Being is tha Sutra interpenetrating all existence. This has a striking similarity with the present day idea of string-like objects in contemporary physics. The strings are one dimensional extended objects and they can get extended /expanded during evolution. Indeed, the followers of Vedas deny the atomic theory of universe in the sense that atoms are geometrical points without parts or magnitude that is indivisible. They have urged that something without parts or magnitude if multiplied by infinite number of times, will remain the same. Anything that has no parts will never make something that has parts; any number of zeros added together will not make one single whole number. So, if the atoms are such that they have no parts or magnitude, the creation of the universe is simply impossible out of such atoms. Hence, the concept of self-extension is an important one towards understanding the nature of space-time. Now one can easily find the existence of two main ingredients :

- **The spontaneous fluctuation**
- **The idea of extension (like hazy lump).**

The moment this concept of extension is introduced, the duality (whole and parts) comes out of being:

" Duality is the beginning of relativity. The moment, two things or two ideas start up, relativity begins.....Here is the beginning of knowledge and elementary reasoning. From this we proceed to more complex reasoning on coming across a variety of things....So we see that relativity is the cause of our knowledge and that space is nothing but an idea of relation" [Ramkrishnananda 1907].

This idea of space as relation is the essence of ontological geometry which we have tried to elaborate throughout this monograph. We propose that in the framework of ontological geometry, the order and the measure arise due to spontaneous fluctuation. Space is considered here as basic structure of all possible facts.A new kind of measure emerge from this "whole fact" or " undivided wholeness". It should be pointed out that mathematicians like Wiener et.al. [Wiener 1966] have both implicitly and explicitly worked in terms of Brownian motion as a basic description (not explained as a result of impacts of finer particles). Such an idea would bring a new kind of order and measure. So space-time should be considered as a derived concept. We have already mentioned that random curves or Brownian trajectories may be considered as having indefinitely large degree of order. Then the concept of undivided wholeness can be related to the infinitely large degree of order.

The concept of implicate and explicate orders (as Bohm considered) are nothing but the concept of involution and evolution. During the process of evolution i.e., from gross matter to consciousness, we are moving towards higher and higher degree of order and in case of involutioni.e., from higher mental activities to the matter, the reverse happens. Here, the transition probability from infinitely large degree of order to lesser degree has a different interpretation than that used in deterministic classical dynamical system. In deterministic classical dynamics, the concept of probability is related to the ignorance of the initial conditions and hence the ensemble averages. But in the above framework of ontological geometry, the probability has a different interpretation. It is related to unknowability (intrinsic to the undivided wholeness) rather than ignorance.

To understand this type of conceptual developments, it may be useful if we start with our scientific heritage from the nineteenth century. Nineteenth century is essentially a century of evolution. Two great discoveries in nienteenth century revolutionized our conceptual understanding of nature. The publication of **The Origin of Species** and the first microscopic interpretation to the increase of entropy by Boltzmann are considered as landmarks in the developments of the modern science. Francois Jacob [Jacob 1973], a Nobel laureate biologist beautifully summarized the theory of evolution as :

"The theory of evolution may be summed up essentially into two propositions. First, that all organisms, past, present or future, descend from one or several rare living system which arose spontaneously. Second, that species are derived from one another by natural selection of the best procreation".

Darwin's theory contains two essential ingredients :

- **The idea of spontaneous fluctuations in biological species**
- **The idea of evolution of irreversibility.**

On the other hand, Boltzmann in the kinetic theory of gases introduced the idea that the mechanism of change, of "evolution" can be described in terms of collision between molecules. His basic discovery is the connection between entropy and probability. Both Darwin and Boltzmann considered evolution, probability or randomness as closely related concepts. However, Boltzmann's result is different and even contradictory to that of Darwin. This situation is analyzed in great depth by Prigogine et al. [Prigogine 1984]:

"Probability will reach its maximum when uniformity will be reached. Think about a system formed by two boxes which may communicate through a small hole. Equilibrium will obviously be achieved when the number of particles in the two boxes will be the same. Therefore, the approach to equilibrium corresponds to destruction of privileged initial condition, to forgetting of initial structure, in contrast to Darwin where evolution means relations of new structure....How then to overcome the apparent contradiction between these concepts? We are living in a single universe....life, are inserted in the basic laws, even on the microscopic level....They are basic processes...were considered as deterministic and reversible. Processes involving randomness or irreversibility were considered to be exceptions, near artifacts. Today we see everywhere the role of irreversible process, of fluctuations. The models considered by classical physics appeared to us now to correspond only to limiting situations which can create artificially.... The artificial may be deterministic and reversible. The nature contains essential elements of randomness and irreversibility. This leads to a new view of matter, in which matter is no longer passive as described in the mechanical world view but it associated with spontaneous activity".

The recent neurophysiological findings on spontaneous activities of neurons add new dimension to this debate by considering the relation of neuronal micro events and macro or global brain activities. The view that the physical world is principally indeterministic or intrinsically random may not be such a controversial issue but it may be controversial to state that random processes play a role in the macro activity of the brain (i.e., neuronal spiking patterns). Recent studies of the irregular firing of cortical neurons suggests, however, that this may be the case. Århem [Århem 1996], a neurobiologist pointed out:

"it also seems clear that spontaneous (intrinsic, not externally induced) neuronal activity, irrespective of whether it is random or pseudo random, is fundamental for normal brain function and for the developments of the neuron system during ontogeny (mainly for establishing synaptic contacts)".

So, the spontaneous activity in creation of order or structure or forming pattern might play a significant role not only in ontological theory of space-time or the creation of the universe or evolution of species but also in explaining the brain activities. This kind of change in our conceptual developments in different strata of knowledge as a whole is the beginning of a new dawn and we can really speak about a new dialogue of man with nature.

BIBLIOGRAPHY

References

Amelino-Camelia G.(1997) :"Enlarged bound on the measurability of distances and quantum K-Poincare group" Phys.Lett.B **392**,283.

Århem P. (1996) : "Vertical Information flow in the brain : On Neuronal Microevents and Consciousness" Biosystem, **38**, 191.

Aristotle (350 B.C.) : "Metaphysics" Edited by Smith J.A. and Ros W.D. **VIII** (Oxford University Press, 1928).

Asanov G.S. (1985) : "Finsler Geometry, Relativity and Gauge Theories" (D. Reidel, Dordrecht).

Asanov G.S., Ponomarenko S.P. and Roy, S. (1988) : Finslerian Multidimensionality, Associated Gauge Fields and Stochastic Space-Time" Fort. Der. Phys., **36**, 679.

Aurbach Peña de La (1971) : "Stochastic Theory of Quantum Mechanics of Particles with Spin" J. Math. Phys., **12**, 453.

Aurbach Peña de La (1983) : "Stochastic Process Applied to Physics and other Related Fields" edited by Gomez B.et al.(World Scientific, Singapore).

Bach A.,Blank H. and Francke H.(1985) :"Bose-Einstein Statistics Derived from the Statisitcs of Classical Particles" Lett.Nuovo Cimento **43**,195.

Badurek G., Rauch H. and Summhammer J.(1983) :"Time-Dependent superposition of spinors" Phys.Rev.Lett **51**,10105.

Bandopadhyaya P. and Roy S.(1976) : "Some Remarks on Non-Local Field Theory & Space-Time Quantization" : Int. Journ. Theort. Phys. **15**, 323.

Bartel W. et al. (JADE collaboration, 1980) : "Test of Quantum Electrodynamics at PETRA" Phys. Lett. B, **92**, 206.

Belinskii V.A., Khalatnikov I.M. and Lifchitz E.M. (1982) : "A General Solution of the Einstein equation with a Time Singularity" Advances in Phys., **31**, 639.

Berezhk G.E. (1990) : "Frictional Acceleration of Cosmic Rays" (Preprint, Yakutsk).

Bethe H.L (1947) : "The Electromagnetic shift of Energy Levels" Phy. Rev., **72**, 339.

Bergia S., Cannata F. and Pasini A.(1989) "On the Possibility of Interpreting Quantum Mechanics in terms of Stochastic Metric Fluctuations" Phys.Lett.A **137**, 21-27.

Bergia S., Cannata F. and Pasini A. : "Space-Time Fluctuation and Stochastic Mechanics : Problems and Perspectives" (Preprint, Univ.of Bologna).

Bergia S., Cannata F., Rosales J.L. and Sanchez-Gomez J.L. : "El Papel Dela Gravitation en La Fundamentacion dela Mechanica Quantica : Problems Abiertos Y perspectivas" (Preprint).

Berger C. et al.(PLUTO collaboration, 1980) : "Test of QED in the Reactions $e^- + e^- \rightarrow e^- + e^-$ and $e^+ + e^- \rightarrow \mu^+ + \mu^-$ at CMS Energies from 9.4 to 31.6 GeV" Zeit. für Physik C, **4**, 269.

Bhattacharjee P. and Sigl G. (1994) : "Monopole Annihilation and Highest Energy Cosmic Rays" - FERMILAB-Pub - 94/387-A (Nov).

Black M. (1937) : "Vagueness" Phil. Sci., **4**, 427.

Blandford D.R. : "Particle Acceleration Mechanisms" : A Caltech Golderod Preprint (GRP - 338).

Blokhintsev D.I. (1960) : "Fluctuations of Space-Time Metric" IL Nuovo Cimento **XVI**,

382.

Blokhintsev D.I. (1973) : "Stochastic Space and Non-Local Fields" Teo. Math. Fiz. **17**, 153.

Blokhintsev D.I. (1973) : "Space and Time in the Microworld" (D. Reidel, Dordrecht).

Blokhintsev D.I. (1975) : "The Stochastic spaces" Particles and Nuclei **5** (American Physical Society), p. 242.

Bohm D. and Vigier J.P. (1954) : "Model of the Causal Interpretation of Quantum Theory in Terms of a Fluid with Irregular Fluctuations" Phys. Rev. **96**, 208.

Bohm D. (1960) : "Quantum Theory" (Asia Publishing House, Calcutta).

Bohm D. (1971) : "Quantum Theory as an Indication of a new Order in Physics" Ed. by D'Espagnant B. "Foundations of Quantum Mechanics" Procd. Int. School of Physics, Enrico Fermi Course IL.

Bohr N. and Rosenfield L. (1933) : "Zur Frage der Meβbarkeit der electromagnetischen Feldgroben" Det. Kgl. Danske Videnskabernes Selskab, Math-Fys. Medd. XII, No. 8, Lenin & Munksgaard, Kopenhagen.

Bopp F. and Hagg R. (1950) : "Über die Mögliechkeit von Spinmodellen" Naturforschung **5a**, 644.

Born M (1938) : "A Suggestion for Unifying Quantum Theory and Relativity" Proc.Roy.Soc., London **165A**, 291.

Born M. (1955) : "Continuity, Determination and Reality" Kgl. Danske Videnskab. Selskab. Mat. Fys.Medd. **30**, No. 2 (reprinted in Max Born, Ausgewahlte Abhandlungen, Erster Band, Gottingen), 196-219.

Borzeszkowski H.H.V. and Treder H.J. (1988) : "The meaning of Quantum Gravity" (D. Reidel,Dordrecht).

Borzeszkowski H.H.V. and Roy S. (1992) : "Jordan-Fock Uncertainty Relation and Quantum General Theory of Relativity" Found.Phys.,**22**,1079.

Bracci L., Mezzorani G. and Quarati P. (1983) : "Bounds on a Hypothetical Fundamental length" Phys. Lett.,**133B**,231.

Brillouin l. (1960) : "Wave propagation and Group Velocity" (Academic Press, NY).

Broglie L. de (1935) : "Une Remarque suré Interaction entre la Matiere et le Champ Electromagnetique" C.R. Acad. Sci (Paris) **200**, 361.

Broglie L. de (1954) : "Theorie Generale de Particules a Spin" (Gauther-Villars, Paris).

Broglie L. de (1960) : "Non-linear Wave Mechanics" (Elsevier, Amsterdam).

Broglie L. de and Vigier J.P.(1972) : "Photon mass and new Experimental Results on Longitudinal Displacement of Laser Beams near Total Reflection" Phys. Rev. Lett., **28**, 1001.

Broglie L. de (1963) : "Introduction to the Vigier Theory of Elementary Particles" (Elsevier, Amsterdam, Lond, NY).

Bunge M. (1977) : "Treatise on Basic Philosophy"**3**: Ontology I" (D. Reidel, Dordrecht), p. 276.

Busch P., Garbowski M. and Lahti P.J. (1995) : "Operational Quantum Physics" Lecture notes on Physics, No. 31 (Springer, Berlin).

Calogero F.(1997) :"Cosmic Rigin of quantization" Phys.Lett.A **228**,335.

Carrol S., Press H.W. and Turner L.E. (1992) : "Cosmological constant" Ann. Rev. of Astrophys. and Astr.,**30**,499.

Caubet J.P. (1975) : "Lecture Notes in Mathematics ed. by Dold A. and Eckmann B." (Springer, NY),No.451,113.

Chi X. et al. (1993) : "Cosmic Rays of the Highest Energies" Part. Astrophys & Cosmology" Ed. by M.M. Shapiro et al.,(Kluwer Acad.Pub.Co.) p.35.

Chi X. et al. (1993) : "Cosmic Gamma Rays from Collapsing Cosmic Strings" Astropart. Phys.,**1**,239.

Cowsik R. (1979) : "Evolution of Cassiopeia A" Astrophys.J.,**227**, 856.

Cowsik R. (1980) : "Comments on Stochastic Acceleration of Cosmic Rays" Astrophys.J.,**241**,1195.

Cowsik R. (1985) : "Sporadic Acceleration of Cosmic Rays" Astrophys. J.,**155**,344.
Dantzig D. Van (1935) : "The Fundamental Equation of Electromagntism, Independent of Metric Geometry" Cambridge Phil.Soct.Procd.,**30**,421.
Dasgupta R. and Roy S.1990) :"Quantum Statisitcs,Distinghusihability and Random Trajectories" Phys.Lett.A **149**,63.
Dattamajumder D. (1993) : "Fuzzy Mathematics and Uncertainty Management for Decision making in Science and Soceity" Comp. Sci.& Info.,**23**,1.
Dewdney C., Holland P.R.,Kyprianidis A. and Vigier J.P.(1986) :"Trajectories and Spin Vector orientations in the Causal interpretation of the Pauli equation"(Private communication).
Diosi L. (1987) : "A Universal Master Equation For the Gravitational Violation of Quantum Mechanics" Phy.Lett.A.,**120**,377.
Diosi L. and Lukacs B.(1989) :On the minimum uncertainty of space-time geodesics " Phys.Lett. A **142**,331.
Diosi L. and Lukacs B. (1993) : "Calculation of X-ray Signals from Károlyhazy Hazy Space-Time" Phys.Lett.A.,**181**,366.
Diosi L. and Lukacs B.(1993) :"Károlyhazy quantum space-time generates neutron star density in vacuum" IL Nuovo Cimento B **108**,1419.
Diosi L.(1995) :"Quantum Dynamics with two Planck constants and the semiclassical limit"(quant-ph/950 3023).
Donald H. et. al. (1988) : "Tower Gravity Experiment : Evidence for non-Newtonian Gravity" Phys.Rev.Lett.,**60**,2567.
Dresden M. (1977) : "Statistical Mechanics and Statistical Methods in Theory and Applications" Edited by Landman Uzi (Plenum, NY).
Duda R.O. and Hart P.E. (1973) "Pattern Classification and Scene Analysis" (Wiley, New York).
Eddington A.S. (1923) : "The Mathematical Theory of Relativity" (Cambridge Univ. Press).
Eddington A.S. (1953) "Fundamental Theory" (Cambridge Univ. Press).
Efimov G.V. (1963) : "On the Developments of a Non-Local Quantum Field Theory without Ultra Violet Divergences" J.E.T.P.,**44**,2107.
Efimov G.V. and Ivanov, M.V. (1993) : "The Quark Confinement Model of Hadrons" (IOP, Bristol and Philadelphia).
Eichler D. (1980) : "Basic Inconsistency in Models of Intersteller Cosmic Acceleration" Astrophys.J.,**237**,809.
Einstein A. (1924) : Schweiz. Natur. Gessell. Verhand,**85**,85.
Einstein A. (1921) : "Geometry and Experience" Ideas and Opinions (Rupa and Co., 1991),p.232.
Engels F. (1973) : "Anti-Dühring" In K. Marx and F. Engels works, **20**,(Berlin, Dietz).
Feller W.(1957) :"An introduction to probability theory and its applications" Vol.1 (Wiley,NY).
Fenech Ch. and Vigier J.P. (1996) : "Variation of Local Heat Energy and Local Temperature under Lorentz Transformatin" Phys.Lett.A.,**215**, 247.
Fermi E. (1949) : "On the Origin of the Cosmic Radiation" Phys.Rev., **75**,1169.
Fischback E. et al. (1986) : "The Analysis of the Eötvös Experiment" Phys.Rev.Lett.,**56**,3.
Fokker D.A. (1965) : "Time and Space, Weight and Inertia" (Oxford, Pergamon Press).
Frederick C. (1976) : "Stochastic Space-Time and Quantum Theory" Phys.Rev.D,**13**,3183.
Frechet M. (1906) : "Sur Quelques Points du Calcul Fonctionnel" Rend.Circ. Mat.Palermo,**22**,1274.
Frenkel A. (1977) : "On the Possible Connection between Quantum Meachanics and Gravitation" Quantum Mechanics, A half century later (D.Reidel Dördrecht).
Frenkel A. :"Spontaneous Localization of the Wave Function and Classical Behavior" Found.Phys. **20**,159.

Frenkel A. (1995) : "The Hazy Space-Time of the Karolyhazy model of Quantum Mechanics" Ed. by Ferrers M. & Van der Merwe : Fundamental Problems in Quantum Phys.(Kluwar Academic Press) p.89.

Freedman W.L. et.al. (1994) : "Distance to Virgo Cluster Galaxy M100 from Hubble Space Telescope Observations of Cepheids" Nature,**371**,757.

Frieman J.A. (1993) : "Some Recent Developments in Cosmology : Dark Matter, Inflation and COBE" FERMILAB-Conf. 93/080-A.

Fujii Y. (1988) : "New Phenomenological Analysis based on a Scalar- Vector Model of th Fifth Force" Phys.Lett.B,**202**,246.

Fuli L. (1981) : "An Estimate of the Photon Rest Mass" Nuovo Cim. Lett.,**31**,289.

Ghirardi G.C., Rimini A. and Weber T. (1986) : "Unified Dynamics for Microscopic & Macroscopic System" Phys.Rev.D,**34**,470.

Ghirardi G.C., Pearle P. and Rimini A. (1990) : "Markov Processes in Hilbert Space and Continuous Spontaneous Localization of System of Identical Particles" Phys.Rev.A,**42**,78.

Giler M., Szabelska, B. et al. (1985) : Procd.Int.Cosm.Ray Conf., La Jolla,USA,**3**,234.

Gisin N. and Percival I.C.(1993) :"Quantum State Diffusion , localization and quantum dispersion entropy" J.Phys.A **26**,2333.

Gisin N. and Percival I.C.(1993) :"The quantum state diffusion picture of physical processes" J.Phys. A **26**,2245.

Gomez C. (1982) : "Quantum Geometry and Strings" Phys.Lett.B, **118**, 407.

Green M.B., Schwarz H.J. and Witten E. (1987) : "Superstring Theory, Vol. I and II" (Cambridge Univ. Press, Cambridge).

Greisen K. (1966) : "End to the Cosmic Ray Spectrum ?" Phys.Rev.Lett., **16**,748.

Gross J.D. et al. (1988) : "String Theory beyond the Planck scale" Nucl.Phys. B, **303**,407.

Guerra F. and Ruggiero P. (1973) : 'New Interpretation of the Euclidean Markov Field in the Framework of Physical Minkowski Space-Time" Phys.Rev. Lett.,**31**,1022.

Guerra F. (1981) : "Structural Aspects of Stochastic Mechanics and Stochastic Field Theory" Phys.Rep.,**77**,263.

Haken R. (1968) : "Relativistic Kinetic Equations including Radiation Effects I : Vlasov Approximation" Journ.Math.Phys.,**9**,1116.

Hausdorff F. (1914) : "Grundzuge der Mengenlehre" (Leipzig,Germany).

Hayakawa S. (1969) : "Cosmic Ray Physics" (Wiley Sons, N.Y.).

Hayashida N. et al. (1994) :"Observation of a very energetic cosmic ray well beyond the predicted 2.7 K cut-off in the primary energy spectrum" Phys.Rev.Lett. **73**, 3491.

Hawking S. (1970) : "The Conservation of Matter in General Relativity" Comm.Math.Physics, **18**, 301.

Heisenberg W. (1936) : "Zur Theorie der Hohenstrahlung" Z.Phys., **101**, 533.

Hertz H. : "The Principles of Mechanics" (Dover, NY, 1956).

Ingraham R.L. (1962) : "Stochastic Lorentz Observers and the Divergence in Quantum Field Theory" - IL Nuvoo Cim., **24**,1117.

Ingraham R.L. (1967) : "Renormalization Theory of Quantum Field with a cut-off" (Gordon Beach, New York).

Ingraham R.L. (1985) : "Relativistic Transformation Law of Fields : A Slight Generalization consistent with the Equivalence of all Lorentz Frames" Phys.Rev.D,**31**,796.

Jacob F. (1973) : "The Logic of Life" (Pantheon Books, NY).

Johnson N.L. and Kotz S.(1972) :"Distributions and Statisitcs"(Wiley,NY).

Jokipii J.R. (1979) : "Particle Acceleration Mechanisms in Astrophysics" Ed. Arons J., Max G. and Mckea C.: AIP Conf.Procd. No. **56** , (AIP, NY).

Jordan P. and Fock V. (1930) : "Neue Unbestimmtheitseigenschaften des Electromagnetischem Feldes" Z. Phys. **66**, 206.

Kar G., Sinha M. and Roy S. (1993) : "Non-zero Rest Mass of Photon, Maxwell Equations and Conformal Metric Fluctuation" Int.Journ.Theort.
Phys.,**32**,1052.

Karolyhazy F. (1991) : "The breakdown of the Superposition Principle : Sixty two Years

of Uncertainty" Ed. by Miller A.I. (Plenum, N.Y).
Kirzhnits D.A. (1966) : "A Non-Local Quantum Field Theory" Usp. Fiz. Nauk **90**, 129.
Kirzhnits D.A. and Chechin V.A.(1972) :"Ultra-high energy cosmic rays and a possible generalization of relativistic theory" Sovt.J.Nucl.Phys.**15**,585.
Kramosil I. and Michalik J. (1975) : "Fuzzy Metrics and Statistical Metric-Spaces" - Kybernetica, **11**, 336.
Ktorides C.N., Myung H.C. and Santilli R.M.(1980) : "Elaboration of the Recently Proposed Test of Pauli's Principle under Strong Interactions" Phys.Rev.D,**22**,892.
Kunte S.(1977) : Sankhya A **39**,305.
Kyprianidis A., Sardelis B. and Vigier J.P. (1984): "Causal Non-Local Character of Quantum Statistics" Phys.Lett.A, **100**, 228.
Kyprianidis A.,Roy S. and Vigier J.P.(1987) :"Distinghuishability or Indistinguishability in classical and quatum statisitcs" Phys.Lett.A **119**,333.
Lambert H.J. (1770) : "Letter to Kant" In H.E. Fischer Ed. Briechwechsal von Imm Kant,**3 Vols.**, (Munchen : Georg Muller, 1912).
Landau L.D. (1965) : "Collected Papers of Landau" Ed. by D.Ter Harr (Pergamon, N.Y.).
Landau L.D. and Lifchitz E.M. (1962) : "The Classical Theory of Fields" (Pergamon, Oxford).
Lehnert Bo (1995) : "Total Refraction Process including Waves of an Extended Electromagnetic Theory" Optik **99**, 119.
Lehnert B. and Roy S.(1997) :"Extended Electromagnetic Theory" (in press, World Scientific Publishers, Singapore).
Lehr W. and Park, J.L. (1977) : "A Stochastic derivation of the Klein-Gordon Equation" J.Math.Phys.,**18**,1235.
Leibniz von W.G. (1956) : "Philosophical Papers and Letters" Ed. L.H. Loemaker (Chicago Univ. Press).
Lobachevsky N.I. (1914) : "Geometrical Researches on the Theory of Parallels" Translated by Halsted G.B., (Chicago, Open Courb.Pub.)
Lust D. and Theisen S. (1989) : "Lectures on String Theory" (Springer-Verlag, NY).
Mach E. (1909) : "Die Geschichte und die Wurzel des satzes von der Eohaltung der Arbeit" (2nd Ed. Leipzig).
March Von A. (1934) : "Die Geometrie Kleinster RauneI and II" Z. Physik,**104**,93,161.
March Von A. (1937) : "Zur Grundlegung und Anwendung einer Statistischen Metrik" Z. Phys. **105**, 620.
Markov M.A. (1984) : "Some problems of the Recent Theory of Gravitation" Priroda, No.4, p3.
Mather J.C. et al. (1992) : "Recents Results from COBE" In General Relativity and Gravitation (IOP, Bristol), p151.
Menger K. (1942) : "Statistical Matrics" Proc.Nat.Acad.Sci., USA,**28**,535.
Menger K. (1945) : "Theory of Relativity and Geometry" In Albert Einstein : Philosopher Scientist : Ed. Schilpp P.A.,p459.
Menger K. (1951) : "Probabilistic Geometry" Procd.Nat.Acad.Sci., USA,**37**,226.
Menger K. (1979) : "Selected Papers in Logic and Foundations, Didactics, Economics (D.Reidel,Dördrecht), p225.
Miron Radu and Kawaguchi T. (1991) : "Relativistic Geometric Optics" Int.Journ.Theort. Phys.,**30**,1521.
Namsrai Kh. (1986) : "Non-Local Quantum Field Theory and Stochastic Quantum Mechanics" (D. Reidel, Dördrecht).
Namsrai Kh. (1988) : "Quantum Space-Time, A review" Fort.Der Phys., **36**,479.
Namsrai Kh. (1993) : "Stochastic Strings, Topology, and Space-Time Confinement" Int.Journ.Theort.Phys.,**32**,43.
Namsrai Kh. (1996) : "Universal Charge Distribution and Non-Local Quantum Electrodynamics" Int.Journ.Theort.Phys.,**35**,807.
Namsrai Kh. (1996) : "Extended Electron and Nonlocal Electromagnetic Interaction :

The Perturbation Theory" Int.Journ.Theort.Phys.,**35**,807.
Nariai H. (1974) : "On the Average of a Highly Turbulent Gravito -Hydrodynamic Field in th Hadron Era of the Universe" Prog.Theort.Phys., **52**,1539.
Nariai H. (1975) : "On the Average Effect on Highly Turbulent Gravito-Hydrodynamic Field in the Hadron Era of the Universe II" Prog.Theort. Phys.,**53**,656.
Nelson E. (1967) : "Dynamical Theories of Brownian Motion" (Princeton Univ. Press, New Jersy, Princeton).
Nelson E. (1985) : "Quantum Fluctuations" (Princeton Univ. Press, Princeton, New Jersy).
Nielson B.H. and Picek I. (1983) : "Lorentz Non-Invariants" Nucl. Phys.B, **211**,269.
Oppenheimer J.R. (1962) : "The Flying Trapeze: Three crisis for Physicists" (Oxford Univ. Press, N.Y.).
Pais, A. and Uhlenbeck G.E. (1950) : "On Field Theories and Non-Localised Action" Phys.Rev.,**179**,145.
Pauli W. (1923) : "Über das Thermische Gleichgewicht Zwischen Strahlung und Freien Elektonen" Z.Phys.,**1**,272.
Pearle P.(1989) :"Combining Stochastic dynamical state-vector reduction with spontaneous localization" Phys.Rev. **39**,189.
Peña de La and Cetto M.A. (1995) : "The Quantum Dice, An Introduction to Stochastic Electrodynamics" (Kluwer Academic Publishers, Dördrecht).
Penrose R. (1971) : "Angular Momentum : An Approach to Combinatorial Space-Time" Ed. by Bastin T. : Quantum Theory and Beyond (Cambridge Univ. Press, NY),p151.
Penrose R. and MacCullum M.A.H. (1973) : "Twistor Theory : An approach to the Quantization of Fields and Space-Time" Phys.Rep.C, **63**,241.
Penrose R. (1975) : "Twistor theory, its Aims and achievements in Quantum Gravity" (Oxford Univ. Press, N.Y.),p 268.
Percival I.C.(1994) :"Primary state diffusion" Procd.Roy.Soct.Lond.A **447**,189.
Percival C. Ian (1995) : "Quantum Space-Time Fluctuations and Primary State Diffusion" Procd. Royal Soc. London A,**451**,503 and references there in.
Petroni Cufaro N. et al. (1984) : "Causal Stochastic Interpretation of Fermi Dirac Statistics in terms of Distinguishable Non-Locally Correlated Particles" Phys.Lett. A,**100**,4.
Pohl M. (1993) : "Continuous Reacceleration of Cosmic Rays in Galaxies" : Part. Astro Phys. and Cosmology : Ed. by M.M. Shapiro et al., (Kluwer Academic), p17.
Poincaré H. (1898) : "La Measure du Temps" Rev. Meta Phys. and Morale,**V & VI**).
Poincaré H. (1905) : "Science and Hypothesis (Lancester Science Press).
Polyakov M.A. (1987) : "Gauge Fields and Strings" (Harwood, Newark, NJ).
Prigogine Ilya and Stangers Isabelle (1984) : "Order out of Chaos" (Flemingo).
Prugovecki E. (1984) : "Stochastic Quantum Mechanics and Quantum Space-Time" (D.Reidel, Dördrecht).
Ramakrishnanda Swami (1917) : "Space and Time"- The Message of Eternal Wisdom (Sri Ramkrishna Math, Madras, 1980).
Ramanathan R. (1985) : "Why Quantum Diffusion is Non-dissipative?" Phys.Rev.Lett.,**14**,495.
Ramanathan R. (1986) : "Quantum Mechanics, Stochasticity and Space-Time" Physica Scripta,**34**,365.
Ramaty R. and Murphy J.R. (1987) : "Nuclear Processes and Accelerated Particles in Solar Flares" Space Sci.Rev.,**45**,213.
Randers G. (1941) : "On Asymmetrical Metric in the Four-Space of General Relativity" Phys.Rev.,**59**,195.
Ranganathan D. (1988) : "The Dirac Equation in the de Broglie-Bohm Theory"

Phys.Lett. A,**128**,105.
Rauch et al. (1978) : "Precise Determination of the $4-\pi$ Periodicity Factor of a Spinor Wave Function" Z.Phy.B,**29**,281.
Rauch H. and Suda M.(1995) :"Dephasing in neutron interferometry" App.Phys.B **60**,181.
Rauch H. et al. (1996) : "Measurement and Charachterization of thethree Dimensional-Coherence function in Neutron Interferometry" Phys.Rev. A,**53**, 902.
Rauch H. (1996) "Quantum Optics View of Neutron Interferometry" (Preprint).
Riemann B. (1867) : "Über die Hypothesem, Weiche der Geometric Zee Grunde Liegengl" Abh.Kgl.Ges.Wiss., Göttingen,**13**.
Rob A.A. (1936) : "Geometry of Time and Space" (Cambridge Univ. Press, Cambridge).
Rosales J.L. and Sanchez-Gomez J.L. (1991) : "On the propagation of velocity of Metric Fluctuation" Phys.Lett.A, **156**,17.
Rosales J.L. and Sanchez-Gomez J.L. (1992) : "Non-linear Schrödinger equation coming from the Action of the Particle's Gravitational Field on the Quantum Potential" Phys.Lett.A, **166**,111.
Rosales J.L. and Sanchez-Gomez J.L. (1995) : "A Note on some Physical and Cosmological Implications of Conformal Metric Fluctuations" Phys.Lett.A, **199**, 320.
Rosen G. (1980) : "Space-Time Metric Fluctuation induced by Cosmic Turbulence" IL Nuovo Cim.B,**57**,125.
Rosen N. (1947) : "Statistical Geometry and Fundamental Particles" Phys.Rev.,**72**,298.
Rosen N. (1962) : "Quantum Geometry" Annals of Phys.,**19**,165.
Roy S. (1979) : "Relativistic Brownian Motion and Foundations of Quantum Mechanics" IL Nuovo Cim.B,**51**,29.
Roy S. (1980) : "Relativistic Brownian Motion and the Space-Time Approach to Quantum Mechanics" J.Math.Phys.,**21**,71.
Roy S. (1981) : "Relativistic Brownian Motion and the Theory of Measurement in Quantum Mechanics" IL Nuovo Cim.,**64B**,81.
Roy S. (1986) : "Stochastic Geometry and Origin of Quantum Potential" Phys.Lett. A,**115**,256.
Roy S. (1988) : "$4-\pi$ Periodicity of Spinor Wave Function and Stochastic Geometry" Reports on Math. Phys.,**26**,361.
Roy S. and Ramanathan R. (1988) : "Fifth Force and its Possible Origin" (Preprint, Indian Statistical Institute, Calcutta).
Roy S. (1992) : "Random Zero-Point Field and Quantum Correction to the Metric" Acta Applicandae Math.,**26**,209.
Roy S., Kundu M.K. and Granlund G.H. (1996) : "Uncertainty Relations and Time-Frequency Distribution for Unsharp Observables" Inform.Sci., **89**,193.
Roychowdhury R.K. and Roy S. (1982) : "Stochastic Space-Time and Concept of Potential in Classical and quantum Mechanics" Phys.Rev.D,**26**, 1335.
Roychowdhury R.K. and Roy S. (1987) : "Stochastic Geometry and Perturbation of the Energy levels of the Hydrogen Atom" Phys.Lett. A, **123**,429.
Rueda A. (1990) : "Survey and Examination of an Electromagnetic Vacuum Acceleration Effects and its Astrophysical Consequences" Space Sci. Rev.,**53**,223.
Salecker H. and Wigner E.P.(1958) :"Quantum limitations of the measurement of space-time distances" Phys.Rev. **109**,571.
Sanchez-Gomez J.L. and Rosales J.L. (1995) : "Fundamental Problems in Quantum Theory" Annals, New York, Acad. Sci.,**755**,500.
Santilli R.M. (1983) : "Lie-Isotopic lifting of Unitary Symmetric and of Wigner's Theorem for extended, deformable Particles" Lett.IL Nuovo Cim.,**38**, 509.
Sazanov V.N.(1972) :"Possible mass anisotropy and interaction of ultrahigh energy particles with an electromagnetic field" Sovt.J.Nucl.Phys. **15**,590.
Schlickeiser R. (1989) : "Cosmic Ray Transport and Acceleration,I : Derivation of the Kinetic Equation and Applications to Cosmic Rays in Static Cold Media" Astro-

phys.J.,**336**,243.
Schwartz J.H. (1985) : "Superstring, the First Fifteen Years of Superstring Theory" Vol. I and II (World Scientific, Singapore).
Schweizer B. and Sklar A. (1983) : "Probabilistic Metric-Spaces" (North Holland, Amsterdam).
Serstnev A.N. (1967) : "On the Connection between Probabilistic Metric-Spaces and Spacek-Spaces" Kazan Gos Univ. Varm. Zap.,**127**,120.
Shannon C.E. (1948) : "The Mathematical Theory of Communication" Bell System Tech. J., **27**, 379.
Sinha M. and Roy S. (1985) : "Stochastic Space-Time and the Primary Energy Spectrum of Cosmic Rays at $E_0 \geq 10^{19} eV$" IL. Nuovo Cim.C,**8**, 531.
Sinha M. and Roy S. (1987) : "Space-Time Metric Fluctuation and the Cosmological Model with Λ-constant" IL Nuovo Cim.B,**100**,709.
Sinha M. and Roy S. (1987) : "Vacuum Fluctuation and the Acceleration Mechanicsm in High Energy Astrophysics" Procd. 20th ICRC, OG **8.3-22**, Moscow,USSR.
Skawianowski J.J. (1977) : "The Uncertainty Principle and Foundation of Quantum Mechanics" Ed Price W.C. and Chissick S.S.,(Johm Wiley, London), p.147.
Smolin L. (1986) : "On the Nature of Quantum Fluctuation and the Relation to Gravitation and the Principle of Inertia" Classical and Quantum Gravity,**3**,347.
Smolin L. (1986) : "Quantum Gravity and the Statistical Interpretation of Quantum Mechanics" Int.Journ.Theort.Phys.,**25**,215.
Snyder H.S. (1947) : "Quantized space-time" Phys.Rev.,**71**,38.
Snyder H.S. (1947) : "Electromagnetic Field in Quantized Space-Time" Phys.Rev.,**72**,68.
Spacek A. (1956) : "Note on Karl Menger's Probabilistic Geometry" Czech.Math.J.,**61**,72.
Spacek A. (1960) "Random Metric Spaces" Trans. 2nd Prague Conf. Inform. Theories, Statistics, Decision functions and Random Processes, p.627.
Speake C.C. and Quinn T.J. (1988) : "Search for a Short range, Isospin-Coupling component of the Fifth Force with use of a Beam Balance" Phys.Rev.Lett.,**61**,1340.
Stacey F.D. et al. (1987) : "Geophysics and Gravity" Rev.Mod.Phys., **59**,157.
Stapp H.P. (1992) : "Noise induced Reduction of Wave Packets and Faster than Light Influences" Phys.Rev.A,**46**,68.
Stevens R.R. (1968) : "Metrically Generated Probabilistic Metric Spaces" Fund.Math.,**61**,259.
Steven S.S. (1959) : "Measurement, Psychophysics and Utility in Measurement, Definitions and Theories" Ed by Churchman C.W. and Ratoosh P. (John Wiley, NY).
Synge L.J. (1966) : " Relativity, the general Theory" (North Holland, Amsterdam).
Takano Y. (1961) : "A Singularity of Propagators in Field Theory and the Structure of Space-Time" Prog.Theor.Phys.,**26**,304.
Takano Y. (1967) : "Fluctuation of Space-Time and Elementary Particles,I and II" Prog.Theor.Phys.,**38**,1185.
Tamm I.E. (1965) : "On the Curved Momentum Space" Proc.Int.Conf. on Elementary Particles, Kyoto, Japan, p. 314.
Thielheim K.O. (1991) : "Particle Acceleration by Rotating Magnetized Bodies and the Problem of the Origin of High Energy Cosmic Particles" Nucl. Phys.B,**22**,60.
Toneya T(1989) :"On the interpretation of minimal length in string theories" Mod.Phys.Lett. A **4**,1587.
Uhlenbeck G.E. and Ornstein L.S. (1930) : "On the Theory of the Brownian Motion" Phys.Rev.,**36**,823.
Unturbe J. and Sanchez Gomez, J.L. (1992) : "On the Role of Gravitation in the Possible Breakdown of the Quantum Superposition Principle for Macroscopic Systems" IL Nuovo Cim.B,**107**,211.
Unruh G.W. (1976) : "Notes on Black Hole Evaporation" Phys.Rev.D, **14**,870.
Vigier J.P. (1975) : "Model of Quantum Statistics in terms of a Fluid and Irregular Stochastic Fluctuation Propagating at the Velocity of Light : A Derivation of Nelson's

Equations" Lett.Nuovo.Cim.,**24**,265.
Vigier J.P. (1982) : "Louis de Broglie - Physicist and Thinker" Found. Phys.,**12**,923.
Vigier J.P. (1982) : "Non-locality, Causality and Aether in Quantum Mechanics" Astronom. Nach.,**303**,55.
Vigier J.P. and Roy, S. (1985) : "Rauch's Experiment and Causal Stochastic Interpretation of Quantum Statistics" Hadronic J.Suppl.,1,475.
Vigier J.P. (1990) : "Evidence for Non-Zero Mass Photons Associated with Vacuum-induced Dissipative Red-shift Mechanism" IEEE Trans. on Plasma Sci.,**18**,64.
Völk J.H. (1987) : "Particle Acceleration in Astrophysical Shock Waves" : Procd. 20th Int.Cosmic Ray Conf.,Moscow,USSR,**7**,157.
Vivekananda Swami (1900) : "Collected works" Vol **II** and **VI** (Advaita Ashram, Calcutta).
Wald A. (1953) : "On a Statistical Generalisation of Metric Spaces" Proc.Nat.Acad.Su., USA,**293**,196.
Weinberg S.(1972) :"Gravitation and Cosmology :Principles and applications of the general theory of relativity" (Wiley, NY).
Weiner N., Siegel A., Rankin B. and Martin T.W. (1966) : "Differential Space, Quantum Systems and Prediction" (The MIT Press, Cambridge).
Weyl H.(1952) :"Symmetry" (Princeton Univ.Press, Princeton).
Wheeler J.A. (1962) : "Geometrodynamics" (Academic Press, NY).
Wheeler J.A. (1968) : "Superspace and the Nature of Quantum Geometro-Dynamics" Procd. Physical Sessi. Int. School. Nonlinear Math. Phys. (Munich), Ed by Zabuskey N.J. (Springer Verlag, Berlin).
Whitehead A.N. (1969) : "Process and Reality" (Macmillan, NY).
Witten E. (1996) : "Reflections on the Fate of Space-Time" Phys. Today, April, p. 24.
Woehler K. (1971) : "A Conjecture Regarding Quantum Fluctuation of Gravitation and Elementary Particles as Excitons in a Turbulent Gravitational Field" Annals.Phys.,**64**,631.
Yasue K. (1979) : "Stochastic Quantization : A Review" Int.Journ. Theort.Phys.,**18**,861.
Yukawa H. (1950) : "Quantum Theory of Non-Local Fields I and II" Phys.Rev.,**77**,219,1047.
Yukawa H. (1966) : "Atomistics and the Divisibility of Space and Time" Suppl.Prog.Theor.Phys.,No 37 and 38, p512.
Zadeh L.A. (1965) : "Fuzzy sets" Information and Control, **8**, 338.
Zatsepin G.T. and Kuzmin, V.A. (1966) : "Upper Limit of the Spectrum of Cosmic Rays" JETP Lett., **4**, 78.
Zeldovich B.Y.A. (1981) : "Vacuum Theory : a Possible Solution to the Singularity Problem of Cosmology" Sov. Phys. USPEKHI, **24**, 216.
Zeldovich B.Y.A. and Pitaevskai L.P. (1971) : "On the Possibility of the Creation of Particles by a Classical Gravitational Field" Sov. Phys. USPEKHI,**24**,216.
Zurek H.H. (1981) : "Pointer basis of Quantum Apparatus : Into what Mixture does the Wave Packet collapse ?" Phys.Rev.D,**24**,1516.
Zurek H.H. (1982) : "Environment Induced Superselectation rules" Phys.Rev.D,**26**,1862.

INDEX

Fundamental Theories of Physics

Series Editor: Alwyn van der Merwe, *University of Denver, USA*

1. M. Sachs: *General Relativity and Matter.* A Spinor Field Theory from Fermis to Light-Years. With a Foreword by C. Kilmister. 1982 ISBN 90-277-1381-2
2. G.H. Duffey: *A Development of Quantum Mechanics.* Based on Symmetry Considerations. 1985 ISBN 90-277-1587-4
3. S. Diner, D. Fargue, G. Lochak and F. Selleri (eds.): *The Wave-Particle Dualism.* A Tribute to Louis de Broglie on his 90th Birthday. 1984 ISBN 90-277-1664-1
4. E. Prugovečki: *Stochastic Quantum Mechanics and Quantum Spacetime.* A Consistent Unification of Relativity and Quantum Theory based on Stochastic Spaces. 1984; 2nd printing 1986 ISBN 90-277-1617-X
5. D. Hestenes and G. Sobczyk: *Clifford Algebra to Geometric Calculus.* A Unified Language for Mathematics and Physics. 1984 ISBN 90-277-1673-0; Pb (1987) 90-277-2561-6
6. P. Exner: *Open Quantum Systems and Feynman Integrals.* 1985 ISBN 90-277-1678-1
7. L. Mayants: *The Enigma of Probability and Physics.* 1984 ISBN 90-277-1674-9
8. E. Tocaci: *Relativistic Mechanics, Time and Inertia.* Translated from Romanian. Edited and with a Foreword by C.W. Kilmister. 1985 ISBN 90-277-1769-9
9. B. Bertotti, F. de Felice and A. Pascolini (eds.): *General Relativity and Gravitation.* Proceedings of the 10th International Conference (Padova, Italy, 1983). 1984 ISBN 90-277-1819-9
10. G. Tarozzi and A. van der Merwe (eds.): *Open Questions in Quantum Physics.* 1985 ISBN 90-277-1853-9
11. J.V. Narlikar and T. Padmanabhan: *Gravity, Gauge Theories and Quantum Cosmology.* 1986 ISBN 90-277-1948-9
12. G.S. Asanov: *Finsler Geometry, Relativity and Gauge Theories.* 1985 ISBN 90-277-1960-8
13. K. Namsrai: *Nonlocal Quantum Field Theory and Stochastic Quantum Mechanics.* 1986 ISBN 90-277-2001-0
14. C. Ray Smith and W.T. Grandy, Jr. (eds.): *Maximum-Entropy and Bayesian Methods in Inverse Problems.* Proceedings of the 1st and 2nd International Workshop (Laramie, Wyoming, USA). 1985 ISBN 90-277-2074-6
15. D. Hestenes: *New Foundations for Classical Mechanics.* 1986 ISBN 90-277-2090-8; Pb (1987) 90-277-2526-8
16. S.J. Prokhovnik: *Light in Einstein's Universe.* The Role of Energy in Cosmology and Relativity. 1985 ISBN 90-277-2093-2
17. Y.S. Kim and M.E. Noz: *Theory and Applications of the Poincaré Group.* 1986 ISBN 90-277-2141-6
18. M. Sachs: *Quantum Mechanics from General Relativity.* An Approximation for a Theory of Inertia. 1986 ISBN 90-277-2247-1
19. W.T. Grandy, Jr.: *Foundations of Statistical Mechanics.* Vol. I: *Equilibrium Theory.* 1987 ISBN 90-277-2489-X
20. H.-H von Borzeszkowski and H.-J. Treder: *The Meaning of Quantum Gravity.* 1988 ISBN 90-277-2518-7
21. C. Ray Smith and G.J. Erickson (eds.): *Maximum-Entropy and Bayesian Spectral Analysis and Estimation Problems.* Proceedings of the 3rd International Workshop (Laramie, Wyoming, USA, 1983). 1987 ISBN 90-277-2579-9

Fundamental Theories of Physics

22. A.O. Barut and A. van der Merwe (eds.): *Selected Scientific Papers of Alfred Landé. [1888-1975]*. 1988 ISBN 90-277-2594-2
23. W.T. Grandy, Jr.: *Foundations of Statistical Mechanics.* Vol. II: *Nonequilibrium Phenomena.* 1988 ISBN 90-277-2649-3
24. E.I. Bitsakis and C.A. Nicolaides (eds.): *The Concept of Probability.* Proceedings of the Delphi Conference (Delphi, Greece, 1987). 1989 ISBN 90-277-2679-5
25. A. van der Merwe, F. Selleri and G. Tarozzi (eds.): *Microphysical Reality and Quantum Formalism, Vol. 1.* Proceedings of the International Conference (Urbino, Italy, 1985). 1988 ISBN 90-277-2683-3
26. A. van der Merwe, F. Selleri and G. Tarozzi (eds.): *Microphysical Reality and Quantum Formalism, Vol. 2.* Proceedings of the International Conference (Urbino, Italy, 1985). 1988 ISBN 90-277-2684-1
27. I.D. Novikov and V.P. Frolov: *Physics of Black Holes.* 1989 ISBN 90-277-2685-X
28. G. Tarozzi and A. van der Merwe (eds.): *The Nature of Quantum Paradoxes.* Italian Studies in the Foundations and Philosophy of Modern Physics. 1988 ISBN 90-277-2703-1
29. B.R. Iyer, N. Mukunda and C.V. Vishveshwara (eds.): *Gravitation, Gauge Theories and the Early Universe.* 1989 ISBN 90-277-2710-4
30. H. Mark and L. Wood (eds.): *Energy in Physics, War and Peace.* A Festschrift celebrating Edward Teller's 80th Birthday. 1988 ISBN 90-277-2775-9
31. G.J. Erickson and C.R. Smith (eds.): *Maximum-Entropy and Bayesian Methods in Science and Engineering.* Vol. I: *Foundations.* 1988 ISBN 90-277-2793-7
32. G.J. Erickson and C.R. Smith (eds.): *Maximum-Entropy and Bayesian Methods in Science and Engineering.* Vol. II: *Applications.* 1988 ISBN 90-277-2794-5
33. M.E. Noz and Y.S. Kim (eds.): *Special Relativity and Quantum Theory.* A Collection of Papers on the Poincaré Group. 1988 ISBN 90-277-2799-6
34. I.Yu. Kobzarev and Yu.I. Manin: *Elementary Particles. Mathematics, Physics and Philosophy.* 1989 ISBN 0-7923-0098-X
35. F. Selleri: *Quantum Paradoxes and Physical Reality.* 1990 ISBN 0-7923-0253-2
36. J. Skilling (ed.): *Maximum-Entropy and Bayesian Methods.* Proceedings of the 8th International Workshop (Cambridge, UK, 1988). 1989 ISBN 0-7923-0224-9
37. M. Kafatos (ed.): *Bell's Theorem, Quantum Theory and Conceptions of the Universe.* 1989 ISBN 0-7923-0496-9
38. Yu.A. Izyumov and V.N. Syromyatnikov: *Phase Transitions and Crystal Symmetry.* 1990 ISBN 0-7923-0542-6
39. P.F. Fougère (ed.): *Maximum-Entropy and Bayesian Methods.* Proceedings of the 9th International Workshop (Dartmouth, Massachusetts, USA, 1989). 1990 ISBN 0-7923-0928-6
40. L. de Broglie: *Heisenberg's Uncertainties and the Probabilistic Interpretation of Wave Mechanics.* With Critical Notes of the Author. 1990 ISBN 0-7923-0929-4
41. W.T. Grandy, Jr.: *Relativistic Quantum Mechanics of Leptons and Fields.* 1991 ISBN 0-7923-1049-7
42. Yu.L. Klimontovich: *Turbulent Motion and the Structure of Chaos.* A New Approach to the Statistical Theory of Open Systems. 1991 ISBN 0-7923-1114-0

Fundamental Theories of Physics

43. W.T. Grandy, Jr. and L.H. Schick (eds.): *Maximum-Entropy and Bayesian Methods.* Proceedings of the 10th International Workshop (Laramie, Wyoming, USA, 1990). 1991 ISBN 0-7923-1140-X
44. P.Pták and S. Pulmannová: *Orthomodular Structures as Quantum Logics.* Intrinsic Properties, State Space and Probabilistic Topics. 1991 ISBN 0-7923-1207-4
45. D. Hestenes and A. Weingartshofer (eds.): *The Electron.* New Theory and Experiment. 1991 ISBN 0-7923-1356-9
46. P.P.J.M. Schram: *Kinetic Theory of Gases and Plasmas.* 1991 ISBN 0-7923-1392-5
47. A. Micali, R. Boudet and J. Helmstetter (eds.): *Clifford Algebras and their Applications in Mathematical Physics.* 1992 ISBN 0-7923-1623-1
48. E. Prugovečki: *Quantum Geometry.* A Framework for Quantum General Relativity. 1992 ISBN 0-7923-1640-1
49. M.H. Mac Gregor: *The Enigmatic Electron.* 1992 ISBN 0-7923-1982-6
50. C.R. Smith, G.J. Erickson and P.O. Neudorfer (eds.): *Maximum Entropy and Bayesian Methods.* Proceedings of the 11th International Workshop (Seattle, 1991). 1993 ISBN 0-7923-2031-X
51. D.J. Hoekzema: *The Quantum Labyrinth.* 1993 ISBN 0-7923-2066-2
52. Z. Oziewicz, B. Jancewicz and A. Borowiec (eds.): *Spinors, Twistors, Clifford Algebras and Quantum Deformations.* Proceedings of the Second Max Born Symposium (Wrocław, Poland, 1992). 1993 ISBN 0-7923-2251-7
53. A. Mohammad-Djafari and G. Demoment (eds.): *Maximum Entropy and Bayesian Methods.* Proceedings of the 12th International Workshop (Paris, France, 1992). 1993 ISBN 0-7923-2280-0
54. M. Riesz: *Clifford Numbers and Spinors* with Riesz' Private Lectures to E. Folke Bolinder and a Historical Review by Pertti Lounesto. E.F. Bolinder and P. Lounesto (eds.). 1993 ISBN 0-7923-2299-1
55. F. Brackx, R. Delanghe and H. Serras (eds.): *Clifford Algebras and their Applications in Mathematical Physics.* Proceedings of the Third Conference (Deinze, 1993) 1993 ISBN 0-7923-2347-5
56. J.R. Fanchi: *Parametrized Relativistic Quantum Theory.* 1993 ISBN 0-7923-2376-9
57. A. Peres: *Quantum Theory: Concepts and Methods.* 1993 ISBN 0-7923-2549-4
58. P.L. Antonelli, R.S. Ingarden and M. Matsumoto: *The Theory of Sprays and Finsler Spaces with Applications in Physics and Biology.* 1993 ISBN 0-7923-2577-X
59. R. Miron and M. Anastasiei: *The Geometry of Lagrange Spaces: Theory and Applications.* 1994 ISBN 0-7923-2591-5
60. G. Adomian: *Solving Frontier Problems of Physics: The Decomposition Method.* 1994 ISBN 0-7923-2644-X
61 B.S. Kerner and V.V. Osipov: *Autosolitons.* A New Approach to Problems of Self-Organization and Turbulence. 1994 ISBN 0-7923-2816-7
62. G.R. Heidbreder (ed.): *Maximum Entropy and Bayesian Methods.* Proceedings of the 13th International Workshop (Santa Barbara, USA, 1993) 1996 ISBN 0-7923-2851-5
63. J. Peřina, Z. Hradil and B. Jurčo: *Quantum Optics and Fundamentals of Physics.* 1994 ISBN 0-7923-3000-5
64. M. Evans and J.-P. Vigier: *The Enigmatic Photon.* Volume 1: The Field $\boldsymbol{B}^{(3)}$. 1994 ISBN 0-7923-3049-8
65. C.K. Raju: *Time: Towards a Constistent Theory.* 1994 ISBN 0-7923-3103-6
66. A.K.T. Assis: *Weber's Electrodynamics.* 1994 ISBN 0-7923-3137-0

67. Yu. L. Klimontovich: *Statistical Theory of Open Systems*. Volume 1: A Unified Approach to Kinetic Description of Processes in Active Systems. 1995
ISBN 0-7923-3199-0; Pb: ISBN 0-7923-3242-3
68. M. Evans and J.-P. Vigier: *The Enigmatic Photon*. Volume 2: Non-Abelian Electrodynamics. 1995 ISBN 0-7923-3288-1
69. G. Esposito: *Complex General Relativity*. 1995 ISBN 0-7923-3340-3
70. J. Skilling and S. Sibisi (eds.): *Maximum Entropy and Bayesian Methods*. Proceedings of the Fourteenth International Workshop on Maximum Entropy and Bayesian Methods. 1996 ISBN 0-7923-3452-3
71. C. Garola and A. Rossi (eds.): *The Foundations of Quantum Mechanics – Historical Analysis and Open Questions*. 1995 ISBN 0-7923-3480-9
72. A. Peres: *Quantum Theory: Concepts and Methods*. 1995 (see for hardback edition, Vol. 57) ISBN Pb 0-7923-3632-1
73. M. Ferrero and A. van der Merwe (eds.): *Fundamental Problems in Quantum Physics*. 1995 ISBN 0-7923-3670-4
74. F.E. Schroeck, Jr.: *Quantum Mechanics on Phase Space*. 1996 ISBN 0-7923-3794-8
75. L. de la Peña and A.M. Cetto: *The Quantum Dice*. An Introduction to Stochastic Electrodynamics. 1996 ISBN 0-7923-3818-9
76. P.L. Antonelli and R. Miron (eds.): *Lagrange and Finsler Geometry*. Applications to Physics and Biology. 1996 ISBN 0-7923-3873-1
77. M.W. Evans, J.-P. Vigier, S. Roy and S. Jeffers: *The Enigmatic Photon*. Volume 3: Theory and Practice of the $\boldsymbol{B}^{(3)}$ Field. 1996 ISBN 0-7923-4044-2
78. W.G.V. Rosser: *Interpretation of Classical Electromagnetism*. 1996
ISBN 0-7923-4187-2
79. K.M. Hanson and R.N. Silver (eds.): *Maximum Entropy and Bayesian Methods*. 1996
ISBN 0-7923-4311-5
80. S. Jeffers, S. Roy, J.-P. Vigier and G. Hunter (eds.): *The Present Status of the Quantum Theory of Light*. Proceedings of a Symposium in Honour of Jean-Pierre Vigier. 1997
ISBN 0-7923-4337-9
81. M. Ferrero and A. van der Merwe (eds.): *New Developments on Fundamental Problems in Quantum Physics*. 1997 ISBN 0-7923-4374-3
82. R. Miron: *The Geometry of Higher-Order Lagrange Spaces*. Applications to Mechanics and Physics. 1997 ISBN 0-7923-4393-X
83. T. Hakioğlu and A.S. Shumovsky (eds.): *Quantum Optics and the Spectroscopy of Solids*. Concepts and Advances. 1997 ISBN 0-7923-4414-6
84. A. Sitenko and V. Tartakovskii: *Theory of Nucleus*. Nuclear Structure and Nuclear Interaction. 1997 ISBN 0-7923-4423-5
85. G. Esposito, A.Yu. Kamenshchik and G. Pollifrone: *Euclidean Quantum Gravity on Manifolds with Boundary*. 1997 ISBN 0-7923-4472-3
86. R.S. Ingarden, A. Kossakowski and M. Ohya: *Information Dynamics and Open Systems*. Classical and Quantum Approach. 1997 ISBN 0-7923-4473-1
87. K. Nakamura: *Quantum versus Chaos*. Questions Emerging from Mesoscopic Cosmos. 1997 ISBN 0-7923-4557-6
88. B.R. Iyer and C.V. Vishveshwara (eds.): *Geometry, Fields and Cosmology*. Techniques and Applications. 1997 ISBN 0-7923-4725-0

Fundamental Theories of Physics

89. G.A. Martynov: *Classical Statistical Mechanics.* 1997 ISBN 0-7923-4774-9
90. M.W. Evans, J.-P. Vigier, S. Roy and G. Hunter (eds.): *The Enigmatic Photon.* Volume 4: New Directions. 1998 ISBN 0-7923-4826-5
91. M. Rédei: *Quantum Logic in Algebraic Approach.* 1998 ISBN 0-7923-4903-2
92. S. Roy: *Statistical Geometry and Applications to Microphysics and Cosmology.* 1998 ISBN 0-7923-4907-5

KLUWER ACADEMIC PUBLISHERS – DORDRECHT / BOSTON / LONDON

9 780792 349075

GPSR Compliance

The European Union's (EU) General Product Safety Regulation (GPSR) is a set of rules that requires consumer products to be safe and our obligations to ensure this.

If you have any concerns about our products, you can contact us on ProductSafety@springernature.com

In case Publisher is established outside the EU, the EU authorized representative is:

Springer Nature Customer Service Center GmbH
Europaplatz 3
69115 Heidelberg, Germany

Batch number: 10584469

Printed by Printforce, the Netherlands